the darwin Āwařds

EVOLUTION IN ACTION

Wendy Northcutt

A PLUME BOOK

The Darwin Awards: Evolution in Action contains cautionary tales of misadventure.
It is intended to be viewed as a safety manual, not a how-to guide.
The stories illustrate evolution working through natural selection:
Those whose actions have lethal personal consequences
are weeded out of the gene pool. Your decisions can kill you,
so pay attention and stay alive.

For further information about how to avoid the scythe of natural selection,
read Darwin's lessons on safety, science, and the social implications of evolution.

Safety Class
www.DarwinAwards.com/book/teach.html

PLUME
Published by the Penguin Group
Penguin Putnam Inc., 375 Hudson Street, New York, New York 10014, U.S.A.
Penguin Books Ltd, 80 Strand, London WC2R 0RL England
Penguin Books Australia Ltd, Ringwood, Victoria, Australia
Penguin Books Canada Ltd, 10 Alcorn Avenue, Toronto, Ontario, Canada M4V 3B2
Penguin Books (N.Z.) Ltd, 182–190 Wairau Road, Auckland 10, New Zealand

Penguin Books Ltd, Registered Offices: Harmondsworth, Middlesex, England

Published by Plume, a member of Penguin Putnam Inc.
Previously published in a Dutton edition.

First Plume Printing, May 2002
10   9

The Library of Congress has catalogued the Dutton edition as follows:
Northcutt, Wendy.
The Darwin awards : evolution in action / Wendy Northcutt.

p.   cm.
Includes index.
ISBN 0-525-94572-5 (hc.)
ISBN 0-452-28344-2 (pbk.)
1. Stupidity—Anecdotes.   I. Title.
BF431 .N67   2000

081—dc21   00-059634

Printed in the United States of America
Original hardcover design by Leonard Telesca

Praise for *The Darwin Awards*

"Delightfully funny . . . If you are not yet aware of *The Darwin Awards*, you should probably be pitched out of the breeding population . . . Taken together they constitute a delicious sermon in support of common sense."
—*The Baltimore Sun*

"Hilarious . . . A book often is defined as good by saying you can't put it down. With *The Darwin Awards* you can. Then pick it up again. And again."
—*The Flint Journal* (Michigan)

"A warning to all dimwits." —Salon.com

"One of the drawbacks to not teaching the theory of evolution in schools is that some people wind up learning the stuff the hard way . . . Darwin-worthy departures are sent in from people all over the world . . . Fatal stupidity knows no boundaries." —*Sarasota Herald-Tribune*

"D'oh!" —*Creative Loafing* (Atlanta)

WENDY NORTHCUTT is a graduate of UC Berkeley with a degree in molecular biology. She started collecting the stories that make up *The Darwin Awards* in 1993, and founded her award-winning website, www.DarwinAwards.com, soon thereafter. She is the author of *The Darwin Awards II: Unnatural Selection*.

To my sister Elizabeth, who encouraged me to blaze my own path. To my parents, because the apple doesn't fall far from the tree. And to Jacob, who appreciates my oddly successful impetuousness and provided clever chapter titles.

Warm thanks to editor Mitch Hoffman and agent Andrew Stuart, whose expert hands helped shape this book, and to my Philosophy Forum members for their stimulating conversations.

And to Ian.

Whilst this planet has gone on cycling according to the fixed law of gravity, endless forms most beautiful and most wonderful have been, and are being, evolved.

—Charles Darwin in *The Origin of Species*

Contents

CHAPTER 2
Relatively Dangerous: A Family Affair   51

CHAPTER 3
"I Fought the Law . . .": Stupid Criminal Tricks   71

CHAPTER 4
Up in Smoke: Fire and Explosions 103

CHAPTER 5
Leaps of Faith: Fatal Falls 133

CHAPTER 6
Military Intelligence: Uninformed Men   161

CHAPTER 9
Davey Jones' Locker: Watery Demise   229

the darwin Awards

The Darwin Awards:
What Are They?

Darwin Awards illustrate Mark Twain's observation,
"Man is the only animal that blushes—or has reason to."

SURVIVAL OF THE FITTEST

M ost of us know instinctively that the phrase *"trust me, light this fuse"* is a recipe for disaster. Darwin Award winners do not. Most of us have a basic common sense that eliminates the need for public service announcements such as, WARNING: COFFEE IS HOT! Darwin Award winners do not. The stories assembled in this book show that common sense is really not so common.

There are people who think it's practical to peer into a gasoline can using a cigarette lighter. There are people who throw beach parties to celebrate an approaching hurricane. We applaud the predictable demise of such daredevils with Darwin Awards, named after Charles Darwin, the father of evolution. No warning label could have prevented evolution from creeping up on the man who electrocuted fish with household current, then waded in to collect his catch without removing the wire.

Darwin Awards show what happens to people who are bewilderingly unable to cope with obvious dangers in the modern world. The terrorist who mails a letter bomb with insufficient postage wins a Darwin Award when he opens the returned package. As does the fisherman who throws a lit stick of dynamite onto the ice, only to see his faithful golden retriever fetch the stick. As does the man caught stealing from a church.

Darwin Award winners plan and carry out disastrous schemes that an average child can tell are a really bad idea. They contrive to eliminate themselves from the gene pool in such an extraordinarily idiotic manner, that their action ensures the long-term survival of our species, which now contains one less idiot. The single-minded purpose and self-sacrifice of the winners, and the spectacular means by which they snuff themselves, qualifies them for the honor of winning a Darwin Award.

RULES AND ELIGIBILITY

To win, nominees must significantly improve the gene pool by eliminating themselves from the human race in an astonishingly stupid way. All races, cultures, and socioeconomic groups are eligible to compete. Contenders are evaluated using the following five criteria:

The candidate must remove himself from the gene pool.

The prime tenet of the Darwin Awards is that we are celebrating the self-removal of incompetent genetic material from the human race. The potential winner must therefore render himself deceased, or at least incapable of reproducing. If someone does manage to survive an incredibly stupid feat, then his genes *de facto* must have something to offer in the way of luck, agility, or stamina. He is therefore not eligible for a Darwin Award, though sometimes the story is too entertaining to pass up and he earns an Honorable Mention.

Heated philosophical discussions have sprung up around the reproduction rule. If a person or group gives up sex, are they eligible for a nomination since they are no longer willing to breed? Must the candidate be utterly incapable of reproduction? Can the elderly be ruled out because they are too old to have an impact on the gene pool? Should those who already have children be banned from winning?

> Enigmatic philosophical questions: If an identical twin dies in a manner that qualifies him for a Darwin Award, is he still eligible, despite the surviving replica of his genes? Should we logically give Darwins to those who accidentally kill their own children? Suppose a Darwin winner is reincarnated, can he be nominated again in his next life?

These are complicated questions. For example, frozen sperm and ova are viable decades after the donor's demise, and sheep and humans can be cloned from a single cell. It is almost impossible to *completely* eliminate an individual's genes. And it would take a team of researchers to ferret out the full reproductive implications, a luxury the Darwin Awards lacks. Therefore, no attempt is made to determine the actual reproductive status or potential of the nominee. If he no longer has the physical wherewithal to breed with a mate on a deserted island, then he is eligible for a Darwin.

The candidate must exhibit an astounding misapplication of judgment.

We are not talking about common stupidities such as falling asleep with a lit cigarette or taking a bath with a radio. The fatal act must be of such idiotic magnitude that we shake our heads and thank our lucky stars that our descendants won't have to deal with, or heaven forbid breed with, descendants of the buffoon that set that harebrained scheme in motion.

The Darwin winner is seldom a copycat. The death under consideration must reflect a unique manifestation of the grave lack of sense and misapplication of judgment indicative of a genuine cleansing of the gene pool. Using bullets as fuses, reenacting the William Tell stunt, and bungee jumping with rubber bands are all worthy Darwin Award activities.

Oscar Wilde said, "To lose one parent may be regarded as a misfortune . . . to lose both seems like carelessness." If you fry yourself along with your parents while rewiring their outdoor hot-tub during a thunderstorm, you may be eligible for a Darwin Award.

The candidate must be the cause of his own demise.

The candidate's own gross ineptitude must be the cause of the incident that earns him the nomination. A hapless bystander done in by a heavy anvil dropped from a skyscraper is an unfortunate tragedy. If, however, you are smashed by the anvil you rigged above your own balcony to kill those squawking pigeons, then you are a Darwin contender.

A tourist trampled to death by a rampaging bull in a parking lot is merely suffering from bad luck. If you are gored to death during the "running of the bulls" while riding naked in a shopping cart piloted by your drunken friend, you are a candidate for a Darwin Award.

Some feel that a person who intentionally attempts to win a Darwin Award, and succeeds, is by definition a perfect candidate. However, readers should remember that a Darwin Award is an exceedingly dubious honor, and we discourage anyone from intentionally attempting to join these illustrious ranks.

The candidate must be capable of sound judgment.

Humans are generally capable of sound judgment, except those with mental, chemical, or chronological handicaps that render them unable to fully comprehend the ramifications of their actions. That means no children, Alzheimer's disease sufferers, or Downs Syndrome patients. Child nominees are a bone of contention. A vociferous majority argues against letting them win Darwin Awards, citing the gulf between ignorance and stupidity. An equally clamorous minority contends that they are the best candidates for a "rusty chromosome" award, since they obviously have not reproduced. To muddy the ethical waters further, some children have stated that restricting them from vying for this laudable award is yet another encroachment on their civil liberties. We appreciate that parents are responsible for teaching their offspring to make

responsible decisions. Therefore children are not eligible to win a Darwin Award. However, a few are included as nominees, when their actions can be considered foolhardy by even their peers.

The event must be verified.

Reputable newspaper or other published articles, confirmed television reports, and responsible eyewitnesses are considered valid sources. A friend's mother's employer, a chain email, or a doctored photograph are not.

This book contains four categories of stories.

- *Darwin Awards* nominees lost their reproductive capacity by killing or sterilizing themselves, and this is the only category eligible to win a Darwin Award.
- *Honorable Mentions* are foolish misadventures that stop short of the ultimate sacrifice, but still illustrate the innovative spirit of Darwin Award candidates.
- *Urban Legends* are cautionary tales of evolution in action, and are so popular they have become part of the Internet culture. Various versions are widely circulated, but their origins are largely unknown. They should be understood as the fables they are. Any resemblance to actual events, or to persons living or dead, is purely coincidental.
- *Personal Accounts* were submitted by loyal readers blowing the whistle on stupidity, and are plausible

but usually unverified narratives. In some cases readers submitting Personal Accounts have been identified with their permission, but this does not necessarily mean that the sources are directly associated with their Personal Accounts.

Darwin Awards and Honorable Mentions are known or believed to be true. Look for the words *Confirmed by Darwin* under the title, which generally indicate that a story was backed up by multiple submissions and by more than one reputable media source.

Unconfirmed by Darwin indicates fewer credible submissions and the unavailability of direct confirmation of media sources. In "unconfirmed" Darwin Awards or Honorable Mentions, names have often been changed and details of events have been altered to protect the innocent (and for that matter, the guilty).

CHARLES DARWIN'S THEORY OF EVOLUTION

Do the Darwin Awards really represent examples of evolution in action?

In 1859 Charles Darwin revived the theory of evolution in *The Origin of Species*, which presented evidence that species evolve over time to fit their environments better. At that time, the theory of evolution was no longer in vogue. It had already been conceived, discussed, and discredited.

The earth was thought to be only six thousand years old, far too young to show evidence of the slow pace of evolution, and besides, there was no plausible explanation for how evolution might occur. Furthermore, many people were repelled by the notion that man descended from apes. But Darwin's careful biological observations, and his proposed mechanism for evolution, propelled the theory back into the scientific limelight.

Which came first, the chicken or the egg? According to evolutionary theory, the egg did. New species evolve when mutations in parental reproductive cells result in offspring with unique traits. The fertilized egg is the first member of a new species, so the egg comes before the chicken.

Darwin called his mechanism for evolution "natural selection," and described four requirements that must be satisfied in order for natural selection to occur.

First, a species must show variation.

Humans exhibit this quality in abundance. There are variations in every trait you can imagine: height, eye color, emotional balance, toe length, intelligence. We also are very different on the inside. For example, the major artery from the heart may branch either before or after it leaves the left ventricle. Both variations are normal. Your liver may be large or small, your appendix present or absent at birth. Countless differences exist between even the most closely related individuals.

Second, variations must be inheritable.

Children resemble their parents. A staggering number of traits are inherited in the myriad genes we store on our chromosomes. For better or worse, parents pass their genetic strengths and weaknesses on to their offspring. Complex characteristics such as intelligence and personality are influenced by the environment, but even these traits have strong, heritable genetic components.

Third, not all individuals in a population survive to reproduce.

Charles Darwin calculated that a single pair of elephants would multiply to nineteen million in 750 years if each descendant lived 100 years and had six offspring. But the elephant population has remained fairly stable over time. Why aren't we overrun with elephants? Because most of them die without reproducing. As our population boom attests, this criterion is less obviously met by humans; nevertheless, a significant number of people die without reproducing, as the stories in this book show.

Fourth, some individuals can cope with selective pressures better than others.

Due to inherited attributes, some members of a species are more likely to survive predators and cold winters, win the competition for mates, and leave more offspring. Successful traits become more prevalent in the population, while less successful ones decline and eventually die

out. The tales you will read clearly show differences in our ability to cope with the selective pressures that surround us.

Keeping these four criteria in mind, let's follow the example of a hypothetical group of humans with a single variable trait: some are taller than others. Because height is inherited, short people bear shorter children than tall people, on average. Picture these people living in a beautiful

Evolution Outlawed in Kansas?

In August 1999 the Kansas City School District voted to allow references to the theory of evolution to be expunged from science curriculums statewide. Precedent for their anti-Darwin stance is seen in the 1925 prosecution of a Kansas biology instructor for teaching evolution to high-school students.

The Kansas City ruling will probably be reversed by higher courts. The United States Supreme Court ruled in 1987 that requiring schools to teach "creation science" is an unconstitutional endorsement of religion, while requiring schools to teach evolution is not.

"Our school systems teach children that they are nothing but glorified apes who evolutionized out of primordial mud," Texas Representative Tom DeLay declared in a passionate speech on the House floor.

Scientists are bewildered by the fear that evolution continues to inflame in the United States. The mechanisms that Darwin proposed have been reinforced by numerous fossil discoveries, and by trends observed in living species.

But in Kansas community leaders apparently feel that no one in their state has evolved for centuries. And since it's no longer mandatory to include evolutionary theory in their science curriculum, who are we to disagree?

setting among branching trees and scenic cliffs. In this environment, tall people whack their heads on branches and fall over cliffs more frequently than their shorter fellows do. Therefore, short people have a survival advantage, and within a dozen generations, the population will become shorter. It should also become better at evading low branches.

The stories in this book vividly illustrate evolution in all its selective glory, from the sublimely ironic to the pathetically stupid. We think that even Charles Darwin himself would be amused by these examples of trial and fatal error.

UNCOMMON COMMON SENSE

Why are there so many failures of common sense in the modern world?

The world we inhabit today is very different from the world of our ancestors. We evolved to survive on a planet with nothing faster than tigers, and nothing more toxic than broccoli. No carcinogenic man-made chemicals, no explosive fuels or electricity, no refined radioactivity, no mercury thermometers, no lead paint.

Imagine a woman standing in the sun watching squirrels playing in the trees. Imagine that she lives in the past, when there were only a thousand people on earth, and none had thought to smoke tobacco yet. Suddenly, at the speed of light, a photon of ultraviolet radiation travels from the sun to the earth, zaps one of the chromosomes in her

ovary, and changes the sequence of a gene. When that egg becomes an embryo, the result is a child who falls asleep while smoking in bed. He has the Sleepy Smoker gene.

Of course, this is an oversimplification. Complex behaviors don't usually arise from a single mutation. Nevertheless, let's think through the consequences of our hypothetical scenario.

Cigarettes are still unknown in the world, so this child grows up and has children of his own, who also harbor the Sleepy Smoker gene. As the centuries roll by, one in a thousand in our growing population has the dangerous but unexpressed tendency to fall asleep while smoking in bed, and all because one woman's ovary was pierced by a stray bit of radiation.

Eventually shamans discover tobacco, peace pipes become popular in diplomatic circles, and an occasional religious or political figure dies tragically in bed from a side effect of tobacco use. Even so, there just aren't enough people smoking in the world yet to make the consequences significant. The Sleepy Smoker gene continues to proliferate.

Then, in the 1920s, cigarettes are popularized by Hollywood movies. Over the next few decades smoking gains popularity. Suddenly that one person in a thousand is far more likely to be in a situation where his tendency to doze off while smoking in bed will play a role in evolution. Now there is a selective pressure against this particular gene, and the incidence of Sleepy Smoker disease will begin to decline.

Don't take this scenario to heart, and expect to see changes during your lifetime. Evolution works on a grand timescale. It can take hundreds of thousands of years to eradicate a single unfortunate trait. And if we learn to overcome our addiction and stop smoking, the selective pressures against the Sleepy-Smoker gene will ease, and sleepy smokers will continue to proliferate undetected, hidden by a progressive culture.

HISTORY AND INTERNET CULTURE

The philosophy of the Darwin Awards is a way of life.
The origin of the Darwin Awards lies in the infancy of the Internet itself. Darwin Awards were one of the first email chain letters. A story was born when someone with a flair for journalism would notice an example of natural selection in his own backyard, turn it into an amusing anecdote, and send the story to friends. Friends would email friends would email friends, and those original email chains continue even today. They are fossils from the dawn of the Internet.

Some Darwin Awards are short reports based on a single newspaper clipping, such as the man who slept with a gun (FOOLISH INGENUITY: "Midnight Special"). A few turn out to be clever fictions crafted by sardonic writers not content with mere facts. Surreptitiously hidden among authentic Darwin Awards, these legends are known and loved by a microgeneration of fans. Therefore they remain the winners of record, despite being debunked as indicated in the text.

Darwin winners are determined by a lengthy and subjective process. Nominees are culled from the submissions using the the five rules of death, excellence, self-selection, maturity, and veracity. They are written with an eye toward the evolutionary, and made available for public vote and

The author of the JATO legend (TESTOSTERONE POISONING: "JATO") would enjoy a cult notoriety were his identity known today. However, there are several who claim ownership of the *idea* of strapping a jet engine onto a vehicle. One man says he and his friend tried it out on a railroad cart. His twenty-five-thousand-word essay on the subject is an interesting manual of what *not* to do when your father owns a scrapyard.

Origin of the JATO story? Decide for yourself:
www.DarwinAwards.com/book/rocket.html

comment. Thorny issues are debated in the Philosophy Forum, a process illustrated by the John F. Kennedy Jr. debate (LEAPS OF FAITH).

Discredited nominations are removed, and those that fare poorly in the vote are reevaluated for suitability. Community members who believe a story is misrepresented are encouraged to provide an accurate version of events, and stories earning the disapproval of family or community members may be reassessed and removed from consideration. This continuing process of evaluation and revision is perhaps unique to the Internet culture and is made possible by the constant exchange of information among Darwin's thousands of readers. In this manner errors have been eliminated and the stories published here have benefited from that corrective process. At the same time readers should understand that the Darwin Awards and related stories have been built upon this process of community information exchange and are not the results of official investigation. While Darwin is constantly striving to eliminate errors, readers would be wildly missing the point if they were to treat these stories as gospel rather than as humor.

ADVICE ON READING THE STORIES

These stories aren't meant to be read all at once. Like tasty gourmet jelly beans, the flavors are most appealing when you consume a few at a time. A story that makes you laugh out loud when read fresh, may elicit a mental ho-hum after you've surfeited yourself with a dozen others. For maximum enjoyment, be content with a chapter each day.

Remember that a story that makes you laugh may make another recoil with dismay, and vice versa. Reader polls show that, in my quest to illuminate the evolutionary process, I am usually successful at walking the fine line between humor and horror. If you find that I have erred, please turn the page and enjoy the next selection.

As you explore these gems, I hope that you, too, will find joy in the concept of evolution as it applies to our fellow man.

CHAPTER 1

Natural Selection: Animal Misadventures

"Only two things are infinite—the universe and human stupidity, and I'm not so sure about the universe."
—Albert Einstein,
Scientific Advisor to the Darwin Awards.

CAN ANIMALS WIN DARWIN AWARDS?

The simple answer is no. Darwin Awards commemorate individuals whose deaths improve the human gene pool, not the animal gene pool. But that trifling objection could be countered if the Darwin Awards credo were simply changed to read "Darwin Awards commemorate individuals who improve *their species'* gene pool." Then would an animal be eligible for a Darwin Award?

To win a Darwin, one must first behave stupidly. And the prerequisite to behaving stupidly is to possess intelligence.

Animals can certainly display intelligence. Lassie, the legendary canine, taught us that dogs are sensible enough to dial 911 and summon help in an emergency. And an impressively smart fox was recently shown on a British news story. Pursued by hunters and dogs, it ran across an electrified railway line. Four of the dogs were electrocuted by the live wire, and another ten were killed when a train plowed through the confused pack. The fox escaped.

It is apparent that animals possess a degree of intelligence.

But animals lack the mental capacity to weigh alternatives. What's dumb for a human is not dumb for a dog. If a human stuffed his head into a potato chip bag to scarf the last scraps, we might laugh at his suffocation, but for a dog, the death is just plain sad.

If animals are to win Darwin Awards for their respective species, the triggering events must be appropriate. For instance, when birds fly into "invisible" windows, their mistake is not of Darwinian caliber. But a bird that singles itself out by repeatedly attempting to peck fleas off a cat is a prime target for natural selection.

Animals can be really stupid, even from their own limited perspectives. Chickens get trampled to death in a rush to be the one to drink the water dripping from the ceiling, while abundant water is available all around. A dozen sheep will follow one another, each stopping to gaze down the cliff at the bodies of its buddies before stepping out into space. We can imagine a few sheep and chickens standing back from the scene of the disaster, shaking their heads and clucking in astonishment at the stupidity of their own species.

In their defense, it is anthropomorphic of us to categorize chickens and sheep as "stupid" for their lack of foresight. Indeed, perhaps it is even hypocrisy. We have bred domestic animals for docility, not intelligence. There is evidence that we are the most intelligent species on earth because we systematically eliminated the competition of our intelligent cousins. Furthermore, domestic animals are living in an artificial environment instead of in their natural habitat. Domesticated pets and livestock are prey to dangers undreamt by Nature.

Suicidal Lemmings

A children's story describes a young lemming who wanders around his neighborhood asking, "Why are all the lemmings jumping off a cliff on Friday?" He asks the owl. He asks his father. He asks a cat. He asks everyone he meets, but nobody knows. They tell him, "That's just what lemmings do."

It is common "knowledge" that lemmings will commit mass suicide, by running into the ocean or launching themselves from a cliff, when their population exceeds the maximum sustainable limit.

Suicidal lemmings are sometimes cited as an argument against evolution. If a herd of lemmings leaps from a cliff, there must certainly be a few in the crowd who are reluctant to follow the leader. What kind of lemmings will predominate in the next generation? Nonjumping lemmings, of course! After a few such incidents, only nonjumpers would remain. So evolution is clearly not working on the lemming population.

How could evolution go so badly awry?

The answer is that lemmings do not commit mass suicide when their population grows too large. They migrate, and during the mass migration, a few animals are pushed from a cliff, or mistake open ocean for a stream. The legend of the suicidal lemming proliferated after the 1958 Disney nature documentary *White Wilderness* showed staged shots of lemmings jumping from a cliff.

Learn more about lemmings!
www.DarwinAwards.com/book/lemmings.html

We animals are all subject to the same process of evolution. Therefore, each species is eligible for Darwin Awards from its own perspective. But the human version of the Darwin Awards is meant to tickle the human funny bone. Since we can't easily relate to the thought processes of animals, we just aren't amused by their foolish deaths. Therefore, animals are not eligible to win Darwin Awards.

But the *human* animal can and does win, as the following stories attest.

DARWIN AWARD: IN A PIG'S EYE
Confirmed by Darwin
4 JULY 1991

Three Eaton men died from a fatally flawed plan on the evening of July 4. James, Billy, and Ashley were killed after their blue Ford pickup rolled over on Country Road 24. Hogs and alcohol were contributing factors to the accident. "We found several beer cans in and around the scene," said Sheriff Andrew Watson. The driver had a blood alcohol content twice the legal limit.

The events unfolded as follows:

The three men spent the national holiday drinking. Later that evening they were struck with a sudden craving for pork chops. "They were popping off fireworks when Jimmy said they ought to go get some eats," reported Billy's girlfriend, Emma. At 11:00 P.M. they drove ten miles to a pig farm, intent on stealing a hog and satisfying that craving for pork chops.

One of the men scaled the fence and tied the end of a rope to a plump quadruped. The other two men started pulling on the four-hundred-pound beast. The stress of a struggling hog was too much for the six-foot chain link fence, and a fourteen-foot section collapsed loudly, startling the other hogs into a stampede.

"I was asleep when I heard this godawful noise," explained the owner of the farm. "I run out of the house with my shotgun and shot off both barrels in the air, and yelled at them to go get on out."

The friends loaded up their stolen pig in a hurry, tied the rope to the truck, and sped down the county road in excess of ninety miles per hour. Unfortunately they forgot to buckle their seat belts. The pig, on the other hand, was strapped in by its leash.

Three miles down the road, the animal began making a commotion in the back of the pickup truck, causing the vehicle to career wildly. The swerving lurches threw the pig from the back of the truck, and it was dragged along the dirt road for about half a mile.

Distracted by the commotion and impeded by the friction of the pig, the driver hit a soft shoulder and rolled the truck forty feet, ejecting all three men from the vehicle and killing them. The victims were discovered at 5:00 A.M. by a passing motorist.

Police caution motorists to drive sensibly on dirt roads, wear seat belts, and refrain from drinking while driving.

The pig lived.

Reference: *Eaton Express Weekly*

DARWIN AWARD: KILLER WHALE RODEO

1999 Darwin Award Winner
Confirmed by Darwin
6 JULY 1999, FLORIDA

A naked man was found dead on the back of a killer whale at SeaWorld in Orlando, a victim of drowning or hypothermia in the fifty-five-degree water. "There were no obvious signs of trauma. He wasn't chewed or dismembered," the sheriff's office said. The body had scrapes on it, possibly signifying that the victim had been dragged along the bottom of the tank.

Is a man who swims with Orcas worthy of a Darwin Award? Clues from his bizarre history may help us decide.

He was identified as a marijuana-smoking drifter named Daniel.

Hare Krishna priest Paul Seaur shared insights into Daniel's personality, gleaned during his month with the community of six worshipers. He had a great love of nature, writing in his journal and feeding wild birds in the temple garden. However, Daniel had difficulty adjusting to the religion's 4:00 A.M. wake-up time, dietary prohibitions, and abstinence from liquor, drugs, sex, and gambling. He preferred to dodge work and meditate in the chapel listening to heavy metal music.

Daniel unexpectedly announced that he was taking a vow of silence, which puzzled the Hare Krishnas since their religion does not urge its members to be silent. He left

abruptly in the spring, breaking his vow long enough to say, "I want to be free. I want to travel around."

During his travels Daniel left a string of petty offenses throughout South Carolina, Washington, Texas, and Florida. Just days before his death, he had completed a three-day sentence in the Indian River County Jail for stealing a 3Musketeers candy bar from 7-Eleven. He resumed his vow of silence in court. "The suspect could not speak," a Vero Beach officer reported, "so instead he used paper and pen to deny the charge."

Three days after his release, our intrepid stoner gained admittance to SeaWorld and loitered near the whale pools until the 10:00 P.M. closing, evading the twenty-four-hour security. After stripping to his bathing trunks, he scaled a three-foot Plexiglas barrier, crossed a short stone wall, and climbed into Tillikum's frigid enclosure using the steps ringing the eighty-by-one-hundred-foot pool.

An employee spotted Daniel's nude form draped just below Tillikum's dorsal fin at 7:35 A.M. the next morning. His swimming shorts were found elsewhere in the tank. Tillikum apparently tried to remove the shorts with his razor-sharp teeth, the medical examiner said.

The nature lover left few clues about his state of mind when he decided to commune with a carnivore the size of a bus. A joint was nestled inside his pile of clothes, but no admission ticket to SeaWorld could be found. Anonymous park workers made a surprise announcement that this was not the first time Daniel had communed with sea mammals. Two years before, they recalled, he had jumped into the manatee tank, which is filled with warmer water and less aggressive creatures.

Notes about Tillikum the killer whale:

The eight-year-old mammal is the largest killer whale in captivity, at twenty-two feet and eleven thousand pounds. He was appraised at $1.5 million when purchased by Sea-World in 1991, where he joined thirteen other killer whales. He was considered dangerous, as he was never trained for human contact. Biologists say he probably played with Daniel like a toy, without realizing that he was a fragile human being.

This is not Tillikum's first encounter with death. He and two other whales were involved in the drowning of a trainer in Victoria, British Columbia, in 1991. She fell into the whale tank at the SeaLand Marine Park and was dragged beneath the surface to her watery demise.

Tillikum is a fecund marine predator, the sire of four calves born during his breeding stay in Florida. In a comparison between Tillikum and Daniel, it's clear who is higher on the evolutionary scale.

Reference: *Sarasota Herald-Tribune, St. Petersburg Times, Chicago Tribune, Orlando Sentinel,* CNN

DARWIN AWARD: PLAYING WITH CATS
1996 Darwin Award Winner
Confirmed by Darwin
2 JANUARY 1996, INDIA

One man was killed at the Calcutta Zoo, and another mauled, when the pair crossed a moat circling a tiger enclosure to put garlands of flowers around the big cat's neck. The attack triggered panic and a near stampede in the zoo.

Prakesh and Suresh, devotees of the goddess Durga, were drinking when they decided to worship the tiger with their innovative adaptation of a religious New Year's greeting. Shiva, a thirteen-year-old Royal Bengal tiger, was not in favor of the plan. He attacked Suresh when the man tossed floral tribute around his neck. Alarmed, Prakesh kicked the tiger in the face to distract him. The tiger obligingly released Suresh and killed Prakesh instead.

"I was shocked to see two young men weaving about in front of a tiger with garlands in their hands," said witness Rakesh Banerjee. "I saw it all. The tiger turned and jumped on the young man, and within moments, the man's head was dangling."

Reference: *Kunal Sen Gupta*, Calcutta, India

Darwin Award:
Hungry Python Kills Owners
Confirmed by Darwin
11 October 1996, New York

If you let a twelve-foot python roam free in your home, you stand a good chance of qualifying for a Darwin Award.

A man was crushed to death by his pet python after failing to keep the snake properly satiated with food. Grant, nineteen, was found unconscious in a small pool of blood, tightly wrapped in a twelve-foot Burmese python (*Molorus bivattatus*) named Damien. The hungry snake had been fed only a single dead chicken one week prior to the event.

At the time of the attack Grant was preparing to feed Damien a live chicken. Herpetologists speculate that the young man forgot to wash the smell of chicken from his hands. It is also possible that the peckish python simply opted for larger prey. When on the brink of a kill the Burmese python can move with deadly speed, and few creatures can elude its grasp.

Wrapped in python, Grant staggered into the hallway to summon help, where he collapsed. Paramedics summoned the strength of body and mind to uncoil the forty-five-pound, five-inch-thick reptile, and hurl it into an adjacent room. They rushed the victim to the hospital, where he died.

Grant and his brother kept a number of snakes, many uncaged, in their Bronx flat. The dead man's mother,

Carmelita, had tried to persuade her son to abandon his hobby, to no avail. "I begged him to get rid of the python. I even threatened to call the police."

Captain Thomas Kelly, from the Forty-sixth Precinct, said: "It looks accidental. Grant may have suspected that his familiarity with Damien placed him above danger, but a hungry python does not quibble about such niceties."

Damien was fed and caged at an animal control center, where he awaited an uncertain fate.

Reptiles are cold blooded and have a slow metabolism. A meal of one chicken per week is about right for a snake this size. Constrictors kill their prey by suffocation, and the more the prey struggles, the more tightly they grasp it with coils and jaws. In animal care facilities, one animal handler is considered sufficient for six feet of python, but an extra attendant is required for every additional three feet of snake. This particular snake should have been handled by three people, and kept in a designated room.

Reference:
Times of London,
New York Times,
Los Angeles Times

DARWIN AWARD: POISONOUS PETS
Confirmed by Darwin
JUNE 1999, DELAWARE

Fifteen venomous snakes were found in the vicinity of a de-
composed body in Stanton. A neighbor's complaints about
the smell led the discovery of eight rattlesnakes, two cobras,
and the three-day-old corpse.

The forty-five-year-old owner of the reptiles was found
ten feet from the open cage of a young diamondback rattler.
Apparently the man was feeding the snake when he was fatally bitten. Residents of the adjacent apartments were
evacuated by the Delaware Animal Rescue team while a
search was conducted for missing vipers.

Neighbors said they had had no idea that the weird loner
kept poisonous snakes. The SPCA notes that it is legal to
keep deadly snakes, provided you apply for a permit.

Reference: MSNBC.com, *Channel 10 News*

DARWIN AWARD: SNAKE CHARMER?
Unconfirmed by Darwin
MAY 1999, THAILAND

A man known for his snake-catching and -charming skills was called to a neighbor's home for an emergency exorcism of a python, which had invaded their dwelling. The middle-aged man rushed into the house in the northern province of Uttaradit, and emerged shortly thereafter holding the snake victoriously aloft in a burlap sack.

He was walking home with the snake when villagers ran into him, heard his bragging story, and asked to see the python. He pulled the snake from the sack and boldly wrapped it around his neck. The wild python, a five-foot-long coil of solid muscle, constricted around him and began to strangle him.

He screamed for help in vain, for the petrified villagers were afraid to approach the serpent. Within minutes he fell to the ground dead. Local policemen forcibly unwrapped the snake from his neck and placed it in captivity.

Reference: Reuters

The anaconda, related to the boa constrictor, is the largest snake in the world. It grows to thirty-five feet in length and weighs three to four hundred pounds. The following excerpt is purported to be taken from the U.S. Government Peace Corps Manual given to volunteers working in the Amazon jungle. It details what to do if an anaconda attacks you.

1. If you are attacked by an anaconda, do not run. The snake is faster than you are.
2. Lie flat on the ground. Place your arms tightly against your sides, and press your legs against one another.
3. Tuck in your chin.
4. The snake will begin to nudge and climb over your body.
5. Do not panic.
6. After the snake has examined you, it will begin to swallow you from the feet end—always from the feet end. Permit the snake to swallow your feet and ankles. Do not panic!
7. The snake will now begin to suck your legs into its body. You must lie perfectly still. This will take a long time.
8. When the snake has reached your knees slowly and with as little movement as possible, reach down, take your knife and very gently slide it into the side of the snake's mouth between the edge of its mouth and your leg, then suddenly rip upwards, severing the snake's head.
9. Be sure you have your knife.
10. Be sure your knife is sharp.

You should take this advice with a grain of salt. Anacondas, boa constrictors, and pythons swallow from the head down ninety-nine percent of the time, almost always after constricting the prey to death. Even if a snake finds a dead animal it will often constrict it to make sure it is dead before eating it.

DARWIN AWARD: BURMESE PYTHON
Unconfirmed by Darwin
1999, NEVADA

A man was found dead in his Nevada residence, the victim of strangulation by his fifteen-foot Burmese python.

The man was handling his pet when the snake mistook his hand for dinner, clamped its jaws around it, and began constricting around his arm and body in an attempt to quell the spasms of its thrashing prey. As the snake began to engulf his hand and arm, the man instructed his hysterical wife, who was too frightened to approach the snake, to call 911. But the authorities arrived too late. The snake had already constricted around its owner's chest and squeezed him breathless.

Snakes are heavily muscled. A large serpent can constrict so tightly that a man cannot unwind it. Once the snake begins to coil around a body part, the prepared owner uses a lever or a sharp knife to persuade it to abandon its intended course of action. Owners of large snakes generally keep such tools handy.

Snakes are also repelled by liquor. Spraying one in the face with high-proof alcohol is an effective method for disengaging their mouth from unwilling prey. Since most of the idiots who get bitten by their fifteen-foot snakes also have alcohol around, this information could save lives, both human and reptilian.

One wonders whether the snake owner used his free arm to flip the bird to his squeamish wife before he died.

DARWIN AWARD:
WHAT'S NEW, PUSSYCAT?
Confirmed by Darwin
MAY 1999, SPAIN

Two German tourists were enjoying their last day of vacation at Safari Park, a wild game park in Alicante. The Safari Park is a controlled reserve hosting a variety of wild animals living in natural habitats. Visitors driving through the park are cautioned not to open the windows, and to remain within their vehicle at all times. Frequent warning signs are posted in many languages, including German.

While driving through a tiger grotto, Willhelm and his companion parked the car, emerged from it for reasons that are unclear, and locked the doors behind them. They were soon set upon by three Bengal tigers lurking in the nearby brush. The big cats, two males and a female ten to twelve years old, pounced on the unfortunate couple, breaking their necks and quickly silencing their screams.

Security guards rushed to the scene to find the woman beheaded and the man disemboweled.

Reference: Europe Press

DARWIN AWARD: WILD ANIMAL LESSON
Confirmed by Darwin
NOVEMBER 1999, SINGAPORE

A busful of excited children can drive anyone to the brink of madness. Perhaps the actions of one bus driver can be explained by his proximity to a herd of shrieking kids.

Xu, forty-one, was one of thirteen tour drivers hired to escort a school tour through the Shanghai World Animals Park. His bus unexpectedly broke down as the convoy passed through a fenced tiger enclosure. You can imagine the hubbub this would cause among a group of students on a wild animal adventure. Needless to say, the park rules clearly forbid leaving the safety of the vehicle, particularly in the tiger grotto.

I can imagine a circumstance in which such a breakdown would be cause for panic. For instance, if you sneak into the park just before it closes, in a convertible with a flimsy cloth covering, accompanied by a date who is eating a rare steak—then waiting in the vehicle for rescue from the tigers would not be an attractive option. But a bus that is part of a convoy of schoolchildren is not in imminent danger of being abandoned to the tigers. Xu must have realized that help would come swiftly.

But instead of waiting inside, besieged by a clamor of children, he climbed out of the bus and began to repair it. A park manager witnessed the consequent deadly incident, as did the children, who watched in horror while tigers savagely mauled their driver. Their hysterical caterwauling summoned a nearby trainer, who drove the tigers from their victim, but it was too late to save Xu from the deadly consequences of the bites to his neck.

As a consolation prize his death provided a memorable example to the children of the danger of stupidity in action.

Reference: *Singapore Straits Times*

DARWIN AWARD: SMARTER ANIMALS
Confirmed by Darwin
16 AUGUST 1999, GERMANY

Man's best friend?
A hunter from Bad Urach was shot dead by his own dog after he left it with a loaded gun. The fifty-one-year-old man was found sprawled next to his car in the Black Forest. A gun barrel was pointing out the window, and his bereaved dog was howling inside the car. The animal is presumed to have pressed the trigger with its paw, and police have ruled out foul play. Since it happened in a hunting preserve, the dog may elect to have the trophy head mounted on a wall in its doghouse.

1991, NICOSIA, CYPRESS

Under similar circumstances, an Iranian hunter named Ali was shot to death near Tehran by a snake that coiled around his shotgun as he pinned the reptile to the ground. Another hunter reported that the victim tried to catch the snake alive by pressing the butt of his gun behind its head. The snake coiled around the butt and pulled the trigger, shooting Ali in the head.

Reference: Islamic Republic News Agency, *Norfolk* (Virginia) *News*, ABC News, Austria Press, Germany Press, Rueters, Bloomberg News, DPA

DARWIN AWARD:
RUNNING OF THE BULLS
Confirmed by Darwin
14 MAY 2000, FRANCE

A Berlin woman attempting to capture a memorable photo of the running of the bulls in the southern town of Nîmes paid for her stupidity with her life. The sixty-eight-year-old photographer removed a metal safety barricade and walked into the middle of the street, camera to her eye, searching for the best camera angle. She was knocked over by a horse whose startled rider could not stop in time, then trampled by the horse and six rampaging bulls before being rescued from the street. She was flown to a nearby hospital, where she died from her injuries.

Reference: *Hamburger Morgenpost*

DARWIN AWARD: SILENCED BY THE LAMBS
Unconfirmed by Darwin
28 JANUARY 1999, LONDON

A flock of sheep charged a well-meaning British farmer's wife and pushed her over a cliff to her death. The woman, sixty-seven, was charged by dozens of sheep after she brought them a bale of hay on the back of a power bike. The sheep rushed forward and rammed the vehicle, knocking Betty and her bike over the edge of a vacant hundred-foot quarry near Durham, in northeastern England. "I saw the sheep surround the bike. The next thing she was tumbling down the incline," a neighbor told reporters. Her husband is being comforted by friends.

Bleating and babbling,
they fell on his neck with a scream.
Wave upon wave of demented avengers
March cheerfully out of obscurity into the dream.
                            —Pink Floyd, "Sheep"

DARWIN AWARD: HUMAN HITCHING POST
Unconfirmed by Darwin
8 MARCH 2000, NEVADA

A woman we'll call Stephanie, twenty-nine, was working with her young and spirited Arabian horse, which she had won in a lottery the previous year. The animal was only partially trained, and still a bit spooky. Every time Stephanie tried to don its bridle, the horse threw back its head and frustrated her efforts.

Then Stephanie had her brilliant idea.

She tied a rope around the Arabian's head, and fastened the other end around her waist to keep the horse from throwing its head back. That way, both hands were free to fasten the bridle.

But horses are five hundred times stronger than people, according to Deputy Sheriff Lance Modispacher, who reported that the horse spooked again, threw Stephanie off her feet, and began running around its paddock, dragging its erstwhile trainer by the rope around her waist. And the rope was short, so she was trampled right under the horse's feet as it ran.

Her father noticed the commotion and ran to help. Unfortunately his two dogs came with him, and started chasing the horse, nipping at its heels. This did not improve Stephanie's situation. He finally managed to lock the dogs away and fetch a knife from the house. With the help of a neighbor, he chased the horse down and cut the rope, freeing the lacerated lass.

But Stephanie had already spent ten minutes under the hooves of her horse, and she died a few hours later at a local hospital, a victim of internal injuries and head trauma, the result of her lamentable decision to tie herself to a skittish horse.

Honorable Mention: Nine Times a Loser
Confirmed by Darwin
October 1998, Australia

A bloke named Gordon who, by amusing coincidence, hails from Darwin, Australia, lost his arm, the use of his legs, and was revived three times on the operating table after an encounter with a king brown snake, the twenty-first most deadly venomous snake in the world. Gordon said, "I still can't believe my arm's been chopped off just for one snake."

Perhaps nine snakes that each bit him once would be more easily believed than the one snake that bit him nine times.

Gordon, who has admitted he was drunk at the time, had been driving with a friend from Mandorah to Darwin when they saw the snake. He picked it up with his left hand "because I was holding a beer in my right one." The snake bit the web of his hand, but Gordon managed to withstand the pain and put it in a plastic bag. He threw the bag in the back of the car.

Once again quoting Gordon, "For some stupid reason, I stuck my hand back in the bag, and it must have smelled blood, it bit me another eight times." They drove him to a nearby hotel, where Gordon was taken by ambulance to the hospital. His friend tried to keep him conscious by, as Gordon said, "whacking me in the head and pouring beer on me."

Despite his friend's quick action, doctors have said that it will take a long time and a lot of rehabilitation before Gordon regains full muscle control. When he does, we anticipate another Darwin Award attempt.

Reference: *Northern Territory News* (Australia)

HONORABLE MENTION:
REVENGE OF THE GOPHER
Confirmed by Darwin
3 APRIL 1995, CALIFORNIA

Anyone who has watched the movie *Caddyshack* will have a good idea of the resilience of gophers. In the spring of 1995 three employees of the Carroll Fowler Elementary School in Ceres received a gopher in good condition. Their subsequent actions show that they were unfamiliar with the movie in particular, and with the vengeful nature of gophers in general.

One janitor and two maintenance men hauled the gopher into a small janitorial closet and apparently decided to kill it. There is no other plausible reason for spraying cleaning solvent on the gopher.

The solvent was designed to remove gum by freezing it and making it easier to scrape up. Elementary schools have an ongoing need for such solvents. But the gopher was stronger than the gum. Three cans later, it was still alive and kicking.

They paused for a moment of silent reflection, and the janitor lit a cigarette in the fume-filled room. The subsequent explosion injured all three men, and sixteen children were treated for scraped knees.

In the aftermath of the explosion, the persecuted gopher was discovered unharmed, clinging to a wall. He was released back into the wild, where he is expected to enjoy years of free drinks in gopher pubs as he tells the story of his brush with death.

Reference: *Sacramento Bee, Hartford Courant*

HONORABLE MENTION:
WOMAN DISARMED BY TIGER
Confirmed by Darwin
22 MAY 2000, COLORADO

Volunteer lends a hand.
A twenty-eight-year-old wildlife volunteer at the Prairie
Wind Animal Refuge was demonstrating the gentility of a
captive Siberian tiger to visitors when the tiger demonstrated a more familiar trait of its species and ripped her
arm off.

The woman had been a volunteer for two years. When a
group of visitors enquired whether the refuge had problems
with people sticking their hands in the cages, she placed her
arm inside the tiger cage and beckoned a full-grown two-
year-old animal. The tiger, which was new to the facility,
sauntered over and began to lick the woman's hand in an apparent display of affection. When the woman playfully
scratched the tiger's nose, the animal recoiled and closed its
jaws around her hand.

The woman discovered that the tiger did not intend to
return her arm, and pulled away in a panic. Another tour
guide reported that the beast worked its way up her shoulder in two seconds and tore her arm off at the socket. He
said he tried to retrieve the arm, but "the tiger did not want
to give it back."

Though the arm was not found, the remaining woman
was airlifted to a nearby hospital and treated for an accidental

amputation. But the tender-hearted tour guide bore no ill will toward the cat, and begged authorities from her hospital bed not to put down the animal as punishment for for its unsavory meal choice.

Reference: CNN, ABC News, the Associated Press,
KCNC-TV, *Denver Rocky Mountain News*

HONORABLE MENTION:
DON'T MESS WITH MAMA BEAR
Confirmed by Darwin
15 JULY 1999, TENNESSEE

An Alabama man disobeyed a fundamental rule of nature when he videotaped a bear cub frolicking near the road in the Great Smoky Mountains National Park. When the cub became bored and trotted back into the woods, the cameraman followed him through a thicket, only to come face to face with Mama Bear and two more cubs.

Mama naturally chased the intruder, swiping at him but not actually striking him. The cameraman, understandably fleet of foot, fled hastily into some broken tree limbs, which stabbed him in the groin and abdomen. Managing to escape without further injury, he was taken to a local hospital for treatment.

The Park Service will remove dangerous animals from the park, but in this case, they determined that the bear acted only to protect her cubs, and decided to leave her alone. We suggest that it is the cameraman who should instead be removed to a more remote area.

Reference: *Knoxville News-Sentinel*

URBAN LEGEND:
MAN GLUED TO RHINO BUTTOCKS
RUSSIA

A Vermont native found himself in a difficult position while touring the Eagle Rock African Safari Zoo with a group of thespians from St. Petersburg. Ronald went to extremes to demonstrate the power of Crazy Glue, one of America's many marvels, to the Russians. To prove the effectiveness of Crazy Glue he rubbed several ounces of the adhesive onto the palms of his hands and jokingly placed them on the buttocks of a passing rhino.

The rhinoceros, a resident of the zoo for thirteen years, was not initially startled, as it has been part of the petting exhibit since its arrival as a baby. However, once it became aware that it was involuntarily stuck to Ronald, it began to panic and charge wildly about the petting area with Ronald along as an unwitting passenger.

"Sally the Rhino hadn't been feeling well," confided caretaker James Douglass. "She was constipated, and had just been given a laxative when the American played his juvenile prank."

During Sally's tirade a shed wall was gored and two fences destroyed, allowing a number of small animals to escape. Three pygmy goats and one duck were stomped to death. During the stampede and subsequent capture, Sally began to feel the effects of the laxative, showering Ronald repeatedly with over thirty gallons of rhinoceros diarrhea. A team of medics and zoo caretakers were needed to remove his hands from Sally's buttocks. "It was tricky. We had to calm her down while shielding our faces from the pelting rhino dung. I guess you could say that Ronald was in it up to his neck."

Once she was under control, three people with shovels worked to keep an air passage open for Ronald. "We were eventually able to tranquilize Sally and apply a solvent to remove his hands from her rear," said Douglass. "I don't think he'll be playing with Crazy Glue for a while."

Meanwhile, the amused Russians were impressed with the power of the adhesive. "I'm going to buy some for my children, but of course they can't take it to the zoo," commented Vladimir Zolnikov, leader of the troupe.

Ronald did not die, nor was there any reproductive injury, yet he may still qualify for a Darwin Award if you are persuaded by the fact that nobody would date a man who smelled of rhino dung.

RealAudio presentation of Constipated Quadrupeds
www.DarwinAwards.com/book/realaudio1.html

PERSONAL ACCOUNT: BUG REPELLENT
2000, FLORIDA

We moved to Florida when I was in high school in 1972. One summer evening I took a drive across "Alligator Alley," the highway traversing the Everglades. There were so many snakes crisscrossing the pavement, I decided to return the next night to collect a few.

I figured the best approach would be to sit on the car hood while my friend drove me slowly along. When I spied a snake, I'd pound on the hood to alert my pal, and hop off to grab it. Few supplies were needed for this miniexpedition: a pillow sack to hold the snakes and a flashlight to spot them. The mosquitoes would probably be thick again, so I'd bring plenty of repellent too.

My family kept the bug spray in the dark storage area under the kitchen sink, along with many other common household products. In a hurry, I spied the familiar letters O-F-F, grabbed the aerosol canister, and zoomed down to the 'Glades. I couldn't wait to start the snake hunt.

But since I was wearing shorts, I did pause long enough to holler to my friend to toss me the spray. I applied copious amounts to first one thigh, then the other. Then . . .

Yow!

Suddenly the foam of a thousand scrubbing bubbles was frying my skin. I thought I was on fire, but couldn't guess the reason as I ran to the nearby canal and jumped in, hoping a large gator wasn't parked there at the moment. The water eased the pain, and after some intense rubbing I limped back to the car.

My friend was laughing his ass off as he handed me the aerosol can I had dropped in my haste. He shone his flashlight on its label. That industrial-strength bug repellent was so powerful, it would probably have driven off hungry alligators! It was Easy-OFF oven cleaner.

The skin on my thighs eventually sloughed off and healed. Despite this incident, hunting for snakes has been a lifelong passion.

Reference: Bill Love, personal account.

PERSONAL ACCOUNT:
WHY KIDS LEAVE THE FARM
1974, MICHIGAN

Farms in northern Michigan see winter temperatures well below freezing. Materials that normally separate easily, such as grain and water drops, can freeze into solid blocks.

In the winter dairy cows eat silage composed of shredded alfalfa and cornstalks, which is stored in large silos attached to the barns. During one particularly cold snap, the silage froze twelve feet up in a Polish farmer's silo because the top of the silo is more exposed to cold than the lower part adjacent to the barn.

He fed the cows the lower portion of the silage, but the upper portion remained suspended, congealed into a solid roof over the floor of the silo.

Although there was still enough feed for several days, the farmer decided to dislodge the silage. He had some options. He could bring a small heater into the silo to thaw the material overnight. Or he could climb the ladder leading to the top of the silo, and dislodge the ice from a higher vantage. Instead he stepped into the silo and used a long two-by-four to prod the tons of frozen silage above him.

His male children had to use the tractor to extract their father from several tons of silage, so the undertaker could do his job. The obituary listed his death as a "farm accident."

Reference: Brian Bixby, personal account.

CHAPTER 2

Relatively Dangerous: A Family Affair

Frank Zappa observed, "It's not getting any smarter out there. You have to come to terms with stupidity and make it work for you."

CHILDREN AND EVOLUTION

Stories about children that are suggested as Darwin Awards evoke an immediate public outcry. No matter how entertaining the concept of the Darwin Awards is in general, specific examples shown in stories about children will never win approval. What is it about children that makes the average reader blow a fuse?

After all, evolution has *already* shaped our children.

For example, children usually develop a fear of strangers at about the age they learn to walk. At this stage they are extremely attached to their parents, and will become upset if a familiar caretaker is not in sight. Child psychologists think it is an evolutionary adaptation to protect children from harm at a particularly dangerous age. Think how useful that behavior is in a child who has just learned to walk, and can easily wander off into the path of danger without such an attachment.

A multitude of children must have come to an untimely end before this attachment behavior evolved.

Another example is the finicky eating habits children develop as toddlers. The very child who ate nearly everything presented to him at an earlier age, now develops a strong preference for familiar foods and adamantly rejects unfamiliar ones. Even the seemingly innocuous distinction between pureed and sliced food can be significant.

How do you know you don't like beets if you've never tried them?

This pickiness acts as a mechanism to prevent curious toddlers from stuffing poisonous berries and bugs into their throats before they learn which are safe and which are not.

A behavior as complex as that must have evolved through many rounds of natural selection. It is impossible to refute the effects of past evolution on our children today. Evolution has shaped nearly every physical and psychological characteristic we possess.

However, adult behavior has also been subject to evolutionary pressures. Specifically, we have evolved a biological imperative to protect children during an extended childhood. Most of us believe strongly that it is our social responsibility to protect our offspring from harm. And this feeling of responsibility arises in part from genetic programming.

What happens to the species that does not fiercely care for its children? Nothing, if the children can fend for themselves. Fish don't give a second thought to their offspring, yet there are plenty of fish in the sea. But a species that requires parental care during childhood has a biological imperative to protect its young. It is a good thing for humans that we are so protective of our children, and the Darwin Awards will not subvert that programming by poking fun at childish accidents.

However, we *will* poke fun at the accidents suffered by other family members, as you can see from the following stories.

Reader Survey

Does your mother know you read the Darwin Awards?

- **No!**   **67%**
- Yes!   33%

How many members of your family are likely to win a Darwin Award in the next ten years?

- Nobody! We're all OK.   33%
- **One person might win.**   **34%**
- Two or three throwbacks.   14%
- Quite a few of them.   8%
- I will myself!   11%

Would the family members who win a Darwin care if you made their story public for all to read?

- **They'd want to be here.**   **39%**
- They'd be reluctantly willing. 20%
- They would prefer not.   20%
- They would be angry.   21%

The astounding eleven percent of our fans who expect to win a Darwin Award are presumably reading this book for inspiration rather than entertainment. Don't try these tricks at home, or anywhere else!

DARWIN AWARD:
GO, SPEED RACER, GO
Unconfirmed by Darwin
1999, ILLINOIS

Jim had an elderly mother whose driver's license was about
to expire. She didn't want to lose her license, even though
her response time was so slow that the whole family knew
she shouldn't be driving. Jim decided to stop arguing with
his mother, and simply take her to an Illinois driver testing
facility, where she would undoubtedly fail the road test. And
that would be that.

On the fateful day of the test he picked her up and drove
her to the testing facility. It was a warm and sunny day.
When their turn was called, he escorted his mother back to
the car and helped her inside. The driving evaluator began
to climb into the passenger seat.

At this point Jim did something incredibly foolhardy. He
walked directly behind the car, and stood ten feet away
from a brick wall. Standing behind a car is something that
one should avoid even when there is a known *good* driver in
the car. But this car contained a known *bad* driver, and his
actions proved to be rash.

His mother started the engine, stepped on the gas, and
accidentally put the car into gear.

Reverse gear.

The evaluator suffered a broken arm. Poor Jim was not
so lucky. He expired a few days later from internal injuries
sustained in the accident. The Darwin Award goes not only

to Jim, but also to his mother, who avoided passing on her own poor genes by killing her son. After all, Mother does know best.

Reference: Stephen L. Wall, personal account,
Chicago Tribune, Chicago *Daily Herald*

DARWIN AWARD:
WIFE TOSSING IN BUENOS AIRES
1998 Darwin Award Winner
Confirmed by Darwin
FEBRUARY 1998, BUENOS AIRES

Did he win the argument? During a heated marital dispute in a working-class Boedo neighborhood, a twenty-five-year-old man picked up his twenty-year-old wife and threw her off their eighth-floor apartment balcony.

To his dismay she became tangled in the power lines below. He immediately leapt from the balcony and fell toward his wife. We can only speculate as to his reasons. Was he angrily trying to finish the job, or remorsefully hoping to rescue her? He did not accomplish either goal. He missed the power lines completely, and plunged to his death.

The woman managed to swing over to a nearby balcony and was saved.

Reference: Reuters

DARWIN AWARD: COUNT YOUR CHICKENS

1996 Darwin Award Winner
Confirmed by Darwin
31 AUGUST 1995, EGYPT

Six people drowned while trying to rescue a chicken that had fallen into a well in southern Egypt. An eighteen-year-old farmer was the first to descend into the sixty-foot well. He drowned, apparently after an undercurrent in the water pulled him down. Police said his sister and two brothers, none of whom could swim well, went down the well one by one to help him, but also drowned. Two elderly farmers then came by to help. But they were apparently pulled under by the same undercurrent. The bodies of the six were eventually extricated from the well in the village of Nazlat Imara, 240 miles south of Cairo.

The chicken was also pulled out. It survived.

Reference: The Associated Press

RealAudio presentation of "Count Your Chickens"
www.DarwinAwards.com/book/realaudio4.html

DARWIN AWARD: FATHER KNOWS BEST
Confirmed by Darwin
13 MARCH 1999, NEW JERSEY

It started out like a scene from *The Brady Bunch*. Andrew and his fiancée were living together with his three children and her three children in Dover Township, when an argument over chocolate cake icing erupted.

Andrew accused his ten-year-old son of taking the missing container, and the two became embroiled in a heated disagreement. Andrew took the boy out to the garage for a private discussion, and there the conversation became even more emotional. Then the man made his fatal mistake.

He handed a five-inch kitchen knife to his angry son, and challenged the boy to stab him if he hated him so much. The boy put the knife down, but Andrew picked it up and placed it in his hand again. In the heat of the moment the outraged boy took him up on the offer and plunged the knife into his chest. The deadly blow happened so fast that no one could stop it.

Andrew was pronounced dead at Community Medical Center. His last words were "Would you believe the kid did that?"

The fourth grader, charged with manslaughter and illegal possession of a weapon, faces up to three years' imprisonment. But Ocean County prosecutor E. David Millard said it was unlikely that he would serve jail time, as the boy had been provoked.

Reference: Associated Press, *New York Times*,
Philadelphia Daily News, Asbury Park Press

This is classic Darwin. The man will no longer reproduce, and his son obviously had better genes for survival than his father—his mother's contribution, plus the more robust half of his father's genes. Although a man who could inspire such rage in a child is not a sympathetic figure, it is unfortunate that his son was traumatized. One reader suggested tartly that we should invite the boy to accept his late father's trophy. Every person who wins a Darwin Award leaves friends and family members saddened by their loss, but that doesn't lessen their contribution to human evolution.

DARWIN AWARD: LAUGHING GAS
Confirmed by Darwin
16 APRIL 1999, WASHINGTON, D.C.

Considering the manner of their deaths, we can thank our lucky stars that there are two fewer paramedics around. Carol and Mark were found dead in their suburban home by Mark's fourteen-year-old son. The couple were wearing respiratory masks attached to an empty canister of nitrous oxide.

Nitrous oxide, commonly known as "laughing gas," produces a short-lived high, and is often used as a relaxant in dental offices and outpatient clinics. Like every other pure gas, it must be mixed with air or oxygen lest it cause suffocation. Needless to say, Carol and Mark did not mix the nitrous oxide with air.

What makes this story a true Darwin is that both of the deceased had had enough medical training to know better. Mark was a ten-year veteran paramedic with the District of Columbia Fire Department. Carol was studying to become an emergency medical technician for a suburban fire department.

The Washington, D.C., Fire Department's public information officer was quoted as saying Mark was "one of the most educated and highly trained people we had." That must alleviate the concerns of thousands of D.C. residents!

Reference: *Washington Post*

DARWIN AWARD: MURDEROUS AFFAIR

Confirmed by Darwin
Regarding William Padgett. The first article sets the stage,
the second details his innovative way of killing himself.
31 JULY 1878, ENGLAND

William, better know as "Old Bill" Padget, appeared before Justice Brown, charged with attempting to discharge a loaded gun with intent to kill Charles Marshman, for whom he worked upon a farm. The examination showed that on Thursday Bill became angered at Marshman and drew a rifle on him and pulled the trigger, but the cap failed to explode. Marshman struck Bill with a stick of wood, and his fists, and drove him off to the barn, where some parties took the gun away from him and he fled to the woods, where he was found by the officer. Bill is not a very handsome or pleasant looking man when he is all right, and the beating he received had not added to his personal charms. He looked as though he had tempted death by tickling the hind foot of a healthy mule. It was shown that Bill did not know the gun was loaded, he having set it away unloaded, and Marshman had loaded it unbeknown to him. He was held for assault and battery, and on Monday a trial by jury was held. The jury brought in a verdict of "not guilty."

1 FEBRUARY 1887

James and William Padgett were of the first who com-
menced the settlement of this town and voted at this first
election. They settled a few miles from the village near a
stream, which has since been called after them. Bear Trap
Falls on this same stream came by its name in the following
way: A few of their neighbors constructed what is called a
"deadfall" or primitive bear trap, built in the form of a figure
four, with a heavy piece of timber made sharp on one side to
fall upon and hold any large animal when caught under it.
This was in the autumn of 1800. One morning William Padgett while alone examined the trap to see if it was adjusted
correctly. It was, for the sharp log fell and imprisoned the
unfortunate man, and several hours elapsed before anyone
came to his release. He was taken out, called for a drink of
water, which was brought him in a hat from the stream
nearby, when he drank it and immediately expired.

Reference: *Oxford Times*, Reel 9, July 31, 1878,
Oxford Times, Reel 11, February 1, 1887

DARWIN AWARD: AVOIDING A FIGHT
Confirmed by Darwin
19 DECEMBER 1999, NEW YORK

A man died after falling off the roof of a moving car. He was arguing with his girlfriend during a drive home for Christmas along Interstate 88. Although the vehicle was travelling in excess of sixty-five miles per hour in the midst of gale-force winds, the man decided to exit the car onto the roof, presumably to escape from the fight. The luckless boyfriend fell to the ground, where he lay until paramedics rushed him to the hospital. He died the next day due to head injuries. The woman was charged with driving under the influence.

Reference: *Binghamton* (New York) *Press & Sun*

A gentle touch produces soothing neurochemicals. In the same way, an angry shout in a grating voice produces a painful brain state in the target. This man may have been escaping from a pain as real as that produced by physical injury. If his actions were due to irrational responses under the duress of psychic pain, can he truly be considered for a Darwin Award?

Pain and the Darwin Awards
www.DarwinAwards.com/book/pain.html

DARWIN AWARD:
MAINE CHAINSAW ROMANCE
Confirmed by Darwin
5 JULY 1999, MAINE

An Internet romance blossomed then faded, after a Missouri man traveled to Maine to meet his destiny. In a bizarre merging of *You've Got Mail* with *Texas Chainsaw Massacre*, James swung a chainsaw and severed his own neck in a futile effort to prove his love to the woman who had spurned his face-to-face romance.

His relationship with Leigh (not her real name) began over the Internet in 1998, and James moved from Missouri to Maine on June 23 to further the yearlong affair. Instead, Leigh insisted upon ending the relationship post-haste.

Distraught, James drove to her house in Topsham, knocked on the door, and asked her adult son to get his mother. The son refused and locked all the doors. James pulled a chainsaw from his trunk, stood on the lawn, and performed his macho act in a vain attempt to impress the depth of his feelings upon the woman. Police arrived to find him barely alive.

"There was blood all over. I couldn't see where the wound was," explained William Robbins of the Sagadahoc County Sheriff's Department. James died in the hospital shortly thereafter.

A friend of the deceased believes that Leigh abused James's affection. "He spent thousands of dollars on calls, email, computers," she said, "and also helped that woman pay her bills." Debra received a desperate phone call just hours before he took his life. She reported that he begged, "Tell me you forgive me." His friend did so, and then the phone went dead. She attempted to alert authorities but had insufficient information regarding his whereabouts.

James had attempted suicide five years earlier, and had seemingly recovered his equilibrium. He purchased the chainsaw in Maine a week prior to his sensational death.

Reference: *San Jose Mercury News*, Infobeat.com, CNN, *Portland Press*

HONORABLE MENTION:
TRASH COMPACTOR
Unconfirmed by Darwin
14 APRIL 2000, UTAH

A man whose domestic tranquility had been marred by a quarrel with his wife decided to sleep in the relative peace of a garbage Dumpster behind a church. But his private slumber was interrupted on Wednesday morning when he was dumped into a garbage truck and caught in its hydraulic compactor. He was "collected" from behind the church at 6:00 A.M., and the truck proceeded on to gather more rubbish at a high school. The drive had just engaged the truck's compactor when he heard a frantic pounding on the walls of the truck bay. Fire Battalion Chief Brad Wardle commented, "Apparently, he and his wife had an argument. Who knows why he didn't just go to a hotel?"

Reference: *Salt Lake Tribune*

URBAN LEGEND:
MISSIONARY MISCALCULATION
1999, NEW GUINEA

Reverend Upton Down, a self-ordained minister of his own Church of Youthful Wisdom, left Australia with his wife and daughters on a missionary excursion. The ill-fated reverend was determined to bring religion to a tribe of satanic cannibals. He dragged his fearful wife and daughters into the jungles of New Guinea in pursuit of his dream.

His sister begged him not to risk his babies' lives like that. "But he said the cannibals would love his kids, and that would make his job even easier." As it turned out, the Reverend Down guessed wrong. The Tuoari tribesmen promptly ate the Reverend Down, his wife, and his three daughters.

Concerned locals said, "We tried to tell him that Tuoaris don't want missionaries. They are perfectly happy worshiping the devil and eating any juicy white man who comes along."

"Neighboring tribes say that the preacher and his family were in the stewpot before he ever took his Bible out of his duffel bag," Detective Odoka reported. "The Tuoaris ate like kings and danced all night long."

Reference: *Ghana Mirror*

PERSONAL ACCOUNT:
ACCIDENTAL SAFE SEX
1998, MARYLAND

The commanding officer of a student battalion at the Aberdeen Proving Ground was vexed with a student's domestic problem. The man in front of him was one of the many foreign students who came from the Middle East to learn how to fix stuff, but this one was having troubles of a different sort.

He was trying to impregnate his fifteen-year-old wife, who was also his cousin, without success.

The commanding officer arranged for an appointment with a fertility specialist at the Johns Hopkins University Hospital. After examining the couple, the doctor found that both husband and wife were fertile. He then gave them a more extensive interview.

He discovered that the couple had been engaging in a form of intercourse that is not normally associated with procreation. After his careful explanation, they presumably procreated successfully. But offspring or not, they both deserve an Honorable Mention for not knowing how to get the job done.

> Reference: Anonymous personal account, reputed
> to be from the commanding officer.

PERSONAL ACCOUNT: WIVES WITH CHLOROFORM
27 APRIL 1999, WEST VIRGINIA

My great-great-grandfather was the town doctor in Black-water Falls in the 1800s. He lived with his wife, and was affectionately referred to as Dr. Bob by the community.

When he had insomnia, which was frequently, he would get to sleep by the simple expedient of draping a chloroform-soaked handkerchief over his face. His wife had her instructions, and without fail she followed them and removed the cloth once he had fallen asleep.

Unbeknownst to his wife, he was quite affectionate with several women of the area. One day, a teenage girl unexpectedly left a baby boy on his doorstep, claiming it was his and that she could neither care for it nor bear the shame. His wife, grieved that they had never had any children of their own, reluctantly agreed to take the baby in.

That evening the events of the day were unsettling enough to give the doctor severe insomnia. As usual, he draped the chloroformed hanky over his face. However, his wife . . . forgot to remove the handkerchief this time. Old Dr. Bob expired, having pushed his luck too far.

If a Darwin decision like that were a crime, Dr. Bob would be behind bars. In a sense, he is.

The author notes, "The baby was sent to live with Dr. Bob's siblings, and became my great-grandfather, a genius and sire of geniuses. But there's a fine line between genius and nuts, and some of my relatives have crossed that line."

Reference: Janet K. Behning, personal account.

CHAPTER 3

"I Fought the Law . . .": Stupid Criminal Tricks

"I do not approve of anything that tampers with natural ignorance. Ignorance is like a delicate exotic fruit. Touch it and the bloom is gone."
— Lady Bracknell, Royal Patron to the Darwin Awards, in Oscar Wilde's play *The Importance of Being Earnest.*

IF EVOLUTION WORKS, WHY SO MANY IDIOTS?

Look how smart we are! We have the intelligence to completely master our environment. Sure, we evolved from one-celled creatures, but now we are at the top of the food chain. Nothing eats us, everybody can have children, and those children are rarely eaten by sabertooth tigers. We are obviously at the peak of evolution, and we no longer need to evolve." This is a typical argument made by those who accept the theory of evolution, but doubt the existence of *human* evolution.

If humans are no longer evolving, the premise of the Darwin Awards is flawed. And even if the premise of the Darwin Awards is valid, do the particular stories in this book actually represent instances of evolution in action?

The primary difficulty is in imagining genes linked to all the various misguided plans of which we humans conceive. Nobody gives much credence to the idea that there is a gene for smoking in bed. However, smoking in bed may be associated with a variety of heritable traits, such as sensitivity to nicotine addiction. If the daft idea that leads to a Darwin Award has a genetic component, however small, it is a behavior that could potentially be eradicated from the gene pool.

A second reason to doubt whether these stories are illustrations of evolution, is the fact that our environment presents a new set of dangers every few decades. We no longer have to worry much about getting thrown from a horse or being eaten by a bear, but we do need to look both ways before crossing the street. In five hundred years perhaps we will need to avoid ingesting the deadly zsumsu bug, but not have to think twice about being run over by archaic cars, which were replaced by teleporters long ago.

Evolution works on a grand time scale, and the impact of natural selection over mere generations is generally undetectable.

An exception to this rule is when the presence or absence of the gene means certain death. In such circumstances one generation will be sufficient to eliminate all who possess the unfavorable set of chromosomes. Such circumstances are vanishingly rare, but one deadly disease *does* have the potential to create rapid change. That disease is AIDS. The infection rate in sub-Saharan Africa approaches such epidemic levels that it is conceivable that in a few generations, the subcontinent will be inhabited by those few whose genetic makeup makes them less susceptible to AIDS. The implacable hand of evolution is not swayed by human tragedy.

Our own intellect plays a role in evolution, now that many evolutionary pressures are the result of our own inventiveness. We die in car wrecks because we invented the car. We suffer explosion injuries because we invented explosives; we invented explosives because somebody played with sparks; somebody played with sparks because humans are fascinated by fire. In short, we are prone to explosion injuries because our very nature led to the development of explosives.

> Diseases that primarily afflict the elderly take longer for natural selection to eliminate than diseases affecting people with many fecund years ahead of them.

What sort of inherited trait could throttle down such a spiral? Less curiosity would certainly be a start. As you delve into this chapter, ponder an unsettling thought: evolution may be weeding out characteristics of which we humans are particularly fond, such as curiosity!

DARWIN AWARD: JUNK FOOD JUNKIE
1994 Darwin Award Winner
Unconfirmed by Darwin
1994

The 1994 Darwin Award went to the fellow who was killed by a Coke machine, which toppled over on top of him as he was attempting to tip a free soda out of it.

Reference: Reuters, *Morgunbladid* of Iceland, *Kenya Times*

DARWIN AWARD: COPPER CAPER
Confirmed by Darwin
1999, ENGLAND

Wayne wanted to make a few bucks selling stolen scrap metal. He sneaked into a demolition site and surveyed the area for valuable hunks of debris. His eyes fastened upon what appeared to be a three-inch-thick copper pipe. That would fetch a fine fee! But it was too heavy for him to budge. He hauled a few lesser chunks of metal away, but could not get the thought of that copper pipe out of his mind. He returned with sturdy bolt cutters, and it was then, when he attempted to sever the pipe, that he was shocked to discover it was actually carrying eleven thousand volts of power. The paramedics who tried to revive the electrified Wayne were thwarted by the current. He did not survive to be charged with his crime.

Reference: *Derbyshire Times*

DARWIN AWARD:
GOOD TRUMPS EVIL AT CHURCH
Confirmed by Darwin
8 MARCH 1999, KENYA

A middle-aged thief sat quietly through the Sunday service at All Saints Cathedral in Nairobi. But when the offering basket was passed, fellow worshipers were astonished to see him stashing handfuls of the money in his pockets. Realizing he had been spotted, the miscreant fled from the church and onto a busy highway, where a speeding bus killed him. The cause of death? An act of God. The moral? Don't annoy the Ruler of the Universe, or you just may wind up a Darwin Award.

Reader Survey
Do you believe in God?

- **Yes**   55%
- Maybe   24%
- No   21%

DARWIN AWARD: JUMPING JACK CASH
Unconfirmed by Darwin
MARCH 2000, ARIZONA

The Grand Canyon is cordoned off by a fence around the more treacherous overlooks, to prevent unsteady sightseers from tottering to their deaths. Some of these overlooks have small towering plateaus a short distance from the fence. Tourists toss coins onto the plateaus like dry wishing wells. Quite a few coins pile up on the surfaces, while others fall to the valley floor far below.

> Volunteer mountain climbers regularly clean coins from the canyon, to prevent such problems, and donate the proceeds to charity.

One entrepreneur climbed over the fence with a bag, and leapt to one of the precarious, coin-covered perches. He filled the bag with booty, then tried to leap back to the fence with the coins. But the heavy bag arrested his jump, and several tourists were treated to a view of his plunge to the bottom of the Grand Canyon. He did not survive to harvest the piles of coins that had suffered his same fate.

DARWIN AWARD: TIRED OF IT ALL
Confirmed by Darwin
16 AUGUST 1999

Daniel was tired to death—literally—at the Buckeye Ford Dealership in London. He sneaked onto the lot in the wee hours of the morning with theft on his mind. His modus operandi was to jack up the back of a pickup truck, remove the wheels, heave them into the bed of a hot-wired Buckeye Ford pickup, and move on to the next target. Daniel possessed what local police referred to as "an extensive criminal background," and had apparently spent years honing his craft. But his expertise failed him this night. The pickup was half full when the forty-seven-year-old thief's next and final target slipped off the jack and landed squarely on his chest at 4:00 A.M.

A clear case of live by the truck, die by the truck.

Reference: *Columbus Dispatch*, Associated Press

DARWIN AWARD:
MODUS OPERANDI MISFIRES
Unconfirmed by Darwin
1 MARCH 1998, PENNSYLVANIA

Roger, twenty-eight, was a considerate car thief. When the stolen cars became hot, he didn't just abandon them, he torched them. Setting the cars on fire, he reasoned, helped the owners collect insurance on their vehicles. This criminal habit became his downfall. A ten-year career of theft ended when Roger burned to death in Pittsburgh in a van that he had set afire from the inside. He didn't realize that the door handle on the driver's side was broken. His burned body was found inside the van.

DARWIN AWARD: RESTAURANT THIEF
Unconfirmed by Darwin
1992, TENNESSEE

A restaurant in Nashville is well known throughout the music industry, not only for its great food, but also for its star-studded clientele. It is not uncommon for the sidewalk to be littered with long lines of customers waiting for breakfast and the chance to see a famous country star dining there.

The man involved in this story may linger longer in our memories than the average country singer's career.

It seems that one of the employees, noting how successful the place was, thought it would be a perfect place to heist. Early one morning he climbed on the roof and walked to the exhaust chute that hangs over the restaurant's large flat grill. Upon inspection the perpetrator realized that he couldn't negotiate the tight passage fully dressed. If only he hadn't eaten so many free breakfasts! If only he were a few millimeters slimmer! He decided to reduce his bulk by disrobing, and he slid naked down the exhaust chute.

That was the last thing he ever did.

Imagine the surprise of the restaurant opening crew that morning! As they prepared for breakfast, they were horrified to see a pair of legs dangling just inches from the griddle. What happened to our erstwhile villain? It seems that the chute was so tight, there was no room for error. As he slid down the narrow vent, he slipped and caught his own arm under his chin, where he stuck. He died by suffocating himself.

Reference: WSMV TV

Darwin Award: Dum Dum Boutique
Confirmed by Darwin
10 April 1999, New York

Perhaps, as people get older, some really should retire from their careers. Or so it would seem for one fifty-five-year-old burglar. Terrence found new meaning in the term *hanging around late at the bar* when he failed to return home one night. It turned out that he had been breaking and entering through the rooftop window of a shop called the Dum Dum Boutique—the catchy name of a clothing shop—by bending back bars on the window. From this vantage point he made a bold move, and jumped into the store through the gap. Unfortunately, his sweater balked at the sight of all that fashion and refused to join him. It caught on one of the bent bars and strangled him to death. He was found hoist by his own petard the next morning.

References: Reuters, NewsRadio88.com

DARWIN AWARD: SCRAP METAL THIEVES
Confirmed by Darwin
31 JULY 1997

Two teens were disassembling an electric tower with wrenches when it toppled to the ground. They apparently wanted to sell its aluminum supports for scrap, but they failed to realize the essential role the aptly named "support" plays in a 160-foot tower. One of the men was crushed by the collapse of the ten-thousand-pound tower, while the other dug himself out from under, a sadder but wiser man from his close brush with a Darwin Award.

Reference: The Associated Press

Darwin Award:
Wrong Time, Wrong Place

Unconfirmed by Darwin
3 February 1990, Washington

A man tried to commit a robbery in Renton, Washington. It was probably his first attempt at armed robbery, as suggested by the fact that he had no previous record of violent crime, and by his terminally stupid choices:

1. The target was H & J Leather and Firearms. A gun shop.
2. The shop was full of customers—firearms customers—in a U.S. state where a substantial portion of the adult population is licensed to carry concealed handguns in public places.
3. To enter the shop, he had to step around a marked police patrol car parked at the front door.
4. An officer in uniform was standing next to the counter, having coffee before reporting to duty.

Upon seeing the officer, the would-be robber announced a holdup and fired a few wild shots. The officer and a clerk promptly returned fire, covered by several customers who also drew their guns, thereby removing the confused criminal from the gene pool. No one else was hurt.

DARWIN AWARD:
CLUMSY CANADIAN BURGLAR
Confirmed by Darwin
JUNE 1997

A suspected burglar fell to his death from a twelfth-story balcony early yesterday after being surprised by the Calgary apartment's occupants. Residents of the suite are shaken from the incident, and baffled as to how the intruder managed to access their top-floor balcony.

The occupants, a husband and wife, were home at 12:30 A.M. when they heard a noise outside. "We were surprised, but not nearly as surprised as he was," said the husband, whose yell startled the intruder into falling while scrambling to flee. The body of the burglar was found on the ground floor patio directly below the balcony.

This unidentified "cat burglar" lost all nine lives when he failed to land safely on his feet.

Reference: *Calgary Sun*

Darwin Award: Escaping Conviction
Unconfirmed by Darwin
December 1997, Pennsylvania

A prisoner in the new Allegheny County Jail in Pittsburgh attempted to evade his punishment by engineering an escape from his confinement. He constructed a hundred-foot rope of bedsheets, broke through a supposedly shatterproof cell window, and unfurled his makeshift ladder.

It is not known whether his plan took into account the curiosity of drivers on the busy street and Liberty Bridge below. It certainly did not take into account the sharp edges of the glass. By the time he had climbed several feet down the rope, the windowpane had sliced through the cloth and dropped him to his untidy demise. The bottom of the rope was still eighty feet short of the street below.

DARWIN AWARD: ROB YOUR NEIGHBOR
Confirmed by Darwin
25 APRIL 1999, AUSTRALIA

Darren was trying to break into a Craigie neighborhood house as safely and unobtrusively as possible when he wrapped his jacket around his arm and bashed in the window. But the jagged shards tore through the protective cloth and severed an artery in his arm. The thirty-two-year-old stumbled away from the house and through a park, and collapsed eight hundred meters away from the crime scene.

The homeowner returned from a nightclub early that morning to find a broken window, a bloody jacket, and a trail of blood. He searched the jacket and found that it belonged to an acquaintance who he recalled seeing at a tavern on Friday. He telephoned a friend, and they drove to the perpetrator's house to give him a stern reprimand.

When they arrived, they spotted him sleeping in the park nearby. As they approached him, they noted with alarm a trail of blood and his nearly severed arm, and realized that it was too late to lecture him. He had bled to death.

Next time try wrapping your arm in a bulletproof vest instead of flimsy fabric, Darren!

Reference: *West Australia Sunday Times*

This story was originally titled "Next Time Try Kevlar" until a heated debate erupted over the protective merits of Kevlar bullet-proof vests. Apparently, a Kevlar vest will not stop a forceful stabbing with a sharp knife. The point of the knife concentrates all its force on one or two threads. But a knife can stab just as easily through other strong materials, even a car door. Kevlar does offer considerable protection from slashing, comparable to chain mail. Just leaning on large shards of glass is unlikely to penetrate Kevlar.

HONORABLE MENTION: PICK YOUR TARGET
Confirmed by Darwin
19 AUGUST 1999, SPAIN

A professional French pickpocket used astoundingly poor judgment when selecting his most recent victim at the Seville Airport. The thief, who specializes in international events that attract crowds of visitors, thought he was in his element when he circled a group of young men and chose his prey. Little did he realize that he was dipping into the bag of Larry Wade, champion 110-meter hurdler for the U.S. athletic team. He was also spotted by Maurice Green, the fastest sprinter on earth, capable of running 100 meters in 9.79 seconds. The two athletes quickly chased down the thief despite his generous head start. When apprehended, the pickpocket attempted to pretend that he was just an innocent French tourist, but the entire episode was captured on film by a Spanish television crew that had been interviewing Mr. Green at the time. "He chose the wrong man," deadpanned a spokesperson for the Civil Guard.

Reference: the *Times*, the *Times* of London

Honorable Mention: Armed and Dangerous?
Confirmed by Darwin
20 March 2000, Germany

When the masked man stormed into the Volksbank in Niedersachsen and demanded money, the teller complied. Like a child demanding candy, the robber held his bag open with both hands and waited for the cash. Now, any fool knows you can't hold a heavy bag of money and a gun at the same time, so he put the weapon on the counter for a moment. The teller seized his chance and seized the gun, and suddenly the tables were turned. The confused robber raised his arm and, forgetting that his gun was gone, menaced the teller with his index finger. When the robber realized that his situation was not as strong as he had anticipated, he fled the bank on an old bike with pink protection sheet metal. The police are hunting for the man, but they have to take care. He is armed—with his forefinger.

Reference: *Bild am Sonntag*

HONORABLE MENTION:
POOR SENSE OF DIRECTION
Unconfirmed by Darwin
3 DECEMBER 1997, CONNECTICUT

Maurice found himself in custody for making a dreadfully wrong turn, trapping himself in the lobby of a prison as he was fleeing authorities. The confused perp was leading police on a car chase from Suffield and Windsor Locks, when he abruptly pulled into the parking lot of the MacDougall Correctional Institution, a high-security state prison located in Suffield. Maurice leapt from his car and sped into the front lobby, where he was trapped by automatic doors that closed and locked behind him. Police say he apparently thought the building was a shopping mall.

HONORABLE MENTION: AIRBAG WEAPONS
Confirmed by Darwin
APRIL 1999, SOUTH AFRICA

In South Africa carjacking has become popular in recent years. The South African law has lenient provisions for self-defense, and allows "lethal action" if personal property is in danger. Citizens are inventive in creating martial security systems for their cars. Poison gas, acid showers, flame-throwers, and automatic gunfire are not unknown.

One such security system relied upon an airbag installed in the car roof. If a driver sat down without disabling the mechanism, the airbag would inflate and hit the victim atop his head with a force strong enough to render him un-conscious.

And that is exactly what happened to Pieter, who, armed with a pistol, attempted to steal this vehicle. When the airbag exploded, he thought that someone was shooting at him, and he instinctively fired the pistol twice. Unfortu-nately for him, his gun was still in his pocket at the time.

One bullet hit his knee, and the other lodged in the base of his penis.

Reference: *Der Spiegel*

HONORABLE MENTION: MIS-STEAK
Confirmed by Darwin
18 JULY 1999, VIRGINIA

This steak lover will be a "prime" candidate for the Darwin Awards any day now. The story began with a yen for a good steak, and ended behind bars.

Cornelius became embroiled in a dispute with a waffle-house employee over the quality of their steak-and-waffle plate. Police were dispatched to the Fredericksburg, Virginia, Waffle Hut in response to Cornelius's call to 911, crying, "They're taking my money!"

At 1:10 A.M. Sergeant John Barham arrived at the break-fast café and found the man pacing outside the restaurant. The man stated that his order of steak was not properly cooked, and that Waffle Hut had ripped him off by refusing to refund his money. The restaurant manager was interviewed, and agreed to refund the twenty-one dollars to Cornelius.

The diner's victory was short lived. The sergeant ran an identity check, and found that Cornelius was wanted for a probation violation. He hauled him off to the Rappahannock Regional Jail, where he is currently held without bail pending extradition to Florida.

Moral of the story: sometimes moral victories can be decidedly unsatisfying.

Reference: *Fredericksburg* (Virginia) *Free-Lance Star*

Honorable Mention:
Wile E. Coyote of Burglars
Confirmed by Darwin
1 August 1999, California

Myner, twenty-two, broke into a Los Angeles home at 3:00 A.M. on Sunday, only to be confronted by the homeowner, an armed police officer, who fired when he saw the glint of a weapon in the intruder's hand. Myner realized he was in trouble and attempted to flee the scene, but succeeded only in stumbling painfully into a bed of cactus, where he lost his knife. After freeing himself from the prickly plants, he headed toward the fence, a decorative wrought-iron barrier that speared him cruelly in the groin as he hurtled over to the sidewalk. Despite these blunders, he managed to escape, but was apprehended later that morning when he sought treatment for his injuries at the Anaheim Memorial Hospital. Seargent Joe Vargas summed up his adventures by saying, "It wasn't a good night."

Reference: *USA Today, Contra Costa Times*

HONORABLE MENTION:
LIMO AND LATTE BURGLAR
Confirmed by Darwin
1999, WASHINGTON

A penchant for life's little luxuries led to lousy luck for one bungling burglar. This Seattle bank robber rented a limousine, and instructed the chauffeur to drop him off at Bank of America and return when summoned by telephone.

The thief presented a teller with a written demand for money, collected his cash and coins, and ran from the bank to a nearby Starbucks. While he was paying for a double latte with stolen silver, an alert customer phoned police and notified them of the criminal's whereabouts.

While waiting for the latte, the bank burglar placed a call to his chauffeur from a pay phone, and arranged to be picked up outside Starbucks. The police quickly surrounded the store and apprehended the crook, after a brief foot chase, just before his getaway limo arrived. The driver confirmed that he had driven the man to Bank of America.

Reference: *Seattle Times*

Honorable Mention:
Spare Some Change?
Unconfirmed by Darwin
1996, Rhode Island

Portsmouth police charged Garfield, twenty-five, with a string of vending-machine robberies in January. He was captured when he inexplicably fled from police when they spotted him loitering around a favorite vending machine target. Suspicions were confirmed when he tried to post four hundred dollars' bail with four hundred dollars in coins.

HONORABLE MENTION: THREE TIMES A LOSER
Confirmed by Darwin
31 MARCH 2000, NEW MEXICO

Edward had some trouble when he attempted to steal a utility trailer from the Home Depot store in Albuquerque. He
drove in and hitched a trailer onto his Toyota pickup, then
drove off quickly—only to crash on Griegos Road. He then
returned to the home improvement store and hitched up a
second trailer and drove off—only to have it come loose
and crash seventy-five yards away from the first stolen
trailer.

Deputy Sheriff Scott Baird noticed the two trailers on
the side of the road, and stopped to investigate. Just then,
Detective Bill Webb said, Edward "drives by with the third
stolen trailer, and the fender of the trailer clips the deputy's
patrol car." A twenty-five-mile-an-hour chase ensued; the
leisurely pace was set by Edward, who "probably knows that
trailers at high speeds don't stay on very well," Webb
commented.

The would-be thief was finally pulled to a stop, arrested
by Albuquerque police officers, and charged with three
counts of motor vehicle theft.

Three cheers for Edward! If all criminals had a modus
operandi as feeble as his, the species would die out from an
excess of convictions.

Reference: *Albuquerque Journal*

Honorable Mention: Sunny Side Up
Confirmed by Darwin
1999, England

A thief who sneaked into a hospital was scarred for life when he tried to get an artificial suntan. After evading security staff at Odstock Hospital in Salisbury, and helping himself to doctors' paging devices, the thief spotted a vertical sun bed. He walked into the unit and removed his clothes for a forty-five-minute tan. However, the high-voltage UV machine at this hospital, which is renowned for its treatment of burn victims, has a recommended maximum dosage of ten seconds. After lying on the bed for almost three hundred times the maximum safety limit, the man was covered in blisters.

At first, he staunchly bore his pain without complaint, not wanting to return to the hospital he had just burgled. When the pain of the burns became unbearable, he went to Southampton General Hospital, twenty miles away. Staff there became suspicious because he was wearing a doctor's coat, and after tending his wounds they called the police.

Southampton police said: "This man broke into Odstock and decided he fancied a quick suntan." Doctors say he is going to be scarred for life.

Reference: *Times* of London

HONORABLE MENTION: OFFICIAL DRUG TEST
Unconfirmed by Darwin
1997, CANADA

A woman called the police with a complaint that she had been burned in a drug deal. She declared that a man had sold her a rock of crack cocaine, but when she brought it home, it "looked like baking powder." The police dispatched a narcotics agent to her house, who tested the rock and verified that, despite its appearance, it was indeed cocaine. The woman was promptly arrested for drug possession. The RCMP (Royal Canadian Mounted Police) are encouraging anyone who thinks they may have been fooled into buying fake drugs to come forward.

PERSONAL ACCOUNT: KLUTZY CROOK

Unconfirmed by Darwin
FEBRUARY 1998

ATMs have become a popular target for thieves. The law of averages demands that some attempts end unsuccessfully. Our hero knew that in order to collect the prize, he needed to get at the back of a money machine. He pried one away from the wall with difficulty. As soon as he had enough clearance, he wriggled behind it and started working on removing the rear panel. At this point, some problems with his strategy came into play.

He completely ignored the video camera, and apparently did not realize that a silent alarm is triggered if an ATM machine is moved. Furthermore, the ATM in question happened to be only three minutes away from a police station.

As the sirens neared, our novice criminal decided to hide. When the police arrived, they saw only that the machine had been tampered with, and assumed that the perpetrator had fled the scene of the crime. They secured the area and called in a forensics team.

The forensics team was dusting for fingerprints when they heard a very loud sneeze from behind the ATM.

It was not difficult to apprehend the suspect, as he was videotaped, left fingerprints, and chose to hide behind the ATM.

Reference: Sean Barr, personal account, and VOCM Radio

PERSONAL ACCOUNT:
COMPACTED IGNORANCE
MAY 2000, INDIANA

Nurses at the Wabash Valley Correctional Facility in Carlisle have to examine and treat any injuries that occur in the prison during their shift, no matter how outrageous or compromising the offender's situation.

One day the lockdown alarm sounded throughout the prison. An offender was missing, and thought to be an escapee. The nurses drank their coffee and waited for the inevitable capture. An hour later the lockdown ended and they received a call to report to the segregated housing unit, where the escape artist was in need of treatment.

They found him lying on a table, crying, curled in a fetal position.

The offender had crafted what seemed to be a perfect escape. He worked the garbage detail, and he had recruited two other trash collectors to help him escape in the garbage truck. A garbage truck! Who would think to look there?

He asked his two collaborators to bag him up with the trash, load him into the trash compactor, and throw him in the truck with the rest of the rubbish.

Imagine that you are in a maximum-security prison with murderers and rapists. And imagine that you are going to allow, even encourage, two prisoners to seal you into a plastic bag, and put that plastic bag in a very powerful trash compactor. What kind of illogic is that?

The schemer didn't die, but he was a bit squished. His back was never quite the same afterward. If his conspirators hadn't put trash in the bag with him to cushion him, he could have actually won a Darwin Award instead of just an Honorable Mention.

Reference: Anonymous personal account.

PERSONAL ACCOUNT: GANGSTER BLUES
1999, BRAZIL

A car chase in São Paulo was so similar to a Warner Bros. cartoon death that it had the bystanders laughing out loud. A police car was pursuing a car of gangsters, and both began to fire at each other. Suddenly, one of the gangsters had the bright idea to throw a grenade at the police car. He pulled the pin, cocked his arm, and in the heat of the chase, he threw the pin out of the car instead of the grenade. The policemen saw the man doing this, and stopped shooting to watch. The hand grenade exploded in the gangster's car, killing him instantly and wounding the other bandits. The laughter of the citizens, the policemen, and the television anchormen was a paean to natural justice.

Reference: Anonymous personal account.

CHAPTER 4

Up In Smoke:
Fire and Explosions

"The life of man [is] solitary, poor, nasty, brutish, and short."

—Thomas Hobbes

AWARDS FOR PRIESTS AND GAYS?

Should we offer Darwin Awards to groups who have opted out of the opportunity to impact the gene pool by declining to reproduce? There are a number of religious and philosophical reasons to abstain from sex, and groups have perished as a result of such beliefs.

The Shaker movement, a community whose members practiced shaking and trembling to rid themselves of evil, reached its peak of five thousand members in the United States around 1850. But their staunch commitment to celibacy doomed them to dwindle in number. Shaker-run orphanages provided converts for a few generations, but today only one Shaker community remains in America.

Members of the late Heaven's Gate cult took an explicit oath of celibacy. Their book states, "The individual who really recognizes his Heavenly Father doesn't even have the desire to share his heart, soul, and mind with anyone else . . . he is celibate in all ways, not simply in his sensuous nature." They made a brief and fruitless effort to recruit sex addicts to the "Anonymous Sexaholics Celibate Church" on the assumption that those who had already acknowledged their addiction were one step closer to the Heavenly Father. The entire cult perished on March 26, 1997, as the Hale-Bopp comet lit the sky. But even without their mass suicide that day, the cult's stricture against sex guaranteed that its expansion would be self-limiting.

The Voluntary Human Extinction Movement's motto is "May we live long and die out." They encourage a radical alternative to our callous extinction of plants and animals. "Each time another one of us decides to not add another one of us to the burgeoning billions already squatting on this ravaged planet, another ray of hope shines through the gloom."

The philosophies of abstinence held by these three groups guarantee that "the gene stops here."

The Catholic priesthood is a different matter. True, the Catholic clergy do not reproduce. "In the early years of the Church, because of the scarcity of single men who were eligible for ordination, men who were already married were accepted for the priesthood. As the supply of single men became greater, only single men were accepted for ordination, in accordance with Paul's wish that everyone 'be as I am.'" (1 Cor. 7.7)

And yet, the Catholic Church has been around for many years, with no shortage of priests. Catholic clergy exist and flourish because they are spreading memes instead of genes. In a sense, they bud asexually by proselytizing their ideas, which spread virally by continually recruiting new members into the ranks. Catholics are more prodigious than the Shakers because their religious beliefs limit only a minority of the devout to celibacy. Members other than clergy are encouraged to reproduce, and discouraged from using effective birth control. Any religion as successful at enlarging its ranks as the Catholic Church will seldom be lacking in priests.

Homosexuals are arguably eligible for a group Darwin, since they eschew reproductive sex. However, as a plethora of gay and lesbian families attest, they are capable of procreating—using scientific methods when necessary—and are indeed passing their genes on to future generations.

In summary, groups that insist upon celibacy will die off without an effective way of recruiting new converts. And even those celibate groups that manage to expand their ranks will logically eliminate themselves once they reach the hypothetical pinnacle of success: recruitment of the entire human race. These sorts of groups are eligible for a Darwin Award. Catholic priests, homosexuals who manage to reproduce despite their sexual preferences, and similar groups can theoretically exist eternally, and are not eligible for this notorious award.

More information about these groups:
www.DarwinAwards.com/book/group.html

Suppose there were a community of pyromaniacs. How long would such a group survive? The stories in this chapter deal with fire and explosions, and will provide data for answering that question.

DARWIN AWARD: LIVING ON ZIONIST TIME
1999 Darwin Award Winner
Confirmed by Darwin
5 SEPTEMBER 1999, JERUSALEM

The switch away from daylight savings time caused consternation among terrorist groups in 1999.

At precisely 5:30 P.M. Israel Standard Time, two coordinated car bombs exploded in different cities, killing three terrorists who were transporting the bombs. It was initially believed that the devices had been detonated prematurely by klutzy amateurs. A closer look revealed the sardonic truth behind the untimely explosions.

Three days before, Israel had made a premature switch from Daylight Savings Time to Standard Time in order to accommodate a week of Slihot, involving presunrise prayers. Palestinians refused to "live on Zionist time" and kept their clocks on Daylight Savings Time. Two weeks of scheduling havoc ensued.

The bombs were prepared during this unsettled period. They were armed in a Palestine-controlled area, and set on Daylight Savings Time. The confused drivers, however, had already switched to Standard Time. As a result, the cars were still en route when the explosives detonated, delivering the terrorists to their well-deserved demise.

References: *Jerusalem Post*, Associated Press

Darwin Award: Firefighters Ignite!

1999 Darwin Award Winner
Confirmed by Darwin
26 June 1999, Tennessee

Seven firefighters from the Sequoyah Volunteer Fire Department, located in rural Hamilton County north of Chattanooga, decided to impress their chief by surreptitiously setting fire to a house, then heroically extinguishing the blaze. The men allegedly hatched the plan in order to help Daniel, a former firefighter, return to duty.

> See the Honorable Mention "Chimney Safety" on page 118 for more on the history of Snohomish County firefighters.

Unfortunately, Daniel's career plans were irreversibly snuffed when he became trapped while pouring gasoline inside the house. Surrounded by smoke and flames, he was unable to escape, and died inside the burning house.

A reader notes, "What makes me feel this is a genuine candidate is that not only did he kill himself with an act of stupidity, but he is also no longer able to protect other would-be pyromaniacs from Darwin Awards. Had he been successful in his attempt to regain his position, he might have had a ripple effect in the gene pool."

References: Infobeat.com, Associated Press,
Albuquerque Journal, KIRO Radio News

DARWIN AWARD:
IGNITING FIREWORKS THE EASY WAY
Unconfirmed by Darwin
JANUARY 1998, INDONESIA

There are safe methods of lighting fireworks. There are dangerous methods of lighting fireworks. Two residents of villages in East Java were killed when they chose the latter method of ignition.

Firecrackers are illegal in Indonesia. However, they can be purchased on the black market during celebrations such as Eid Al-Fitr, the feast that marks the end of Ramadan. And boys will be boys, the world over.

Prasad, a twenty-eight-year-old resident of Kenongo, and Anan, a twenty-year-old from Telasih, obtained a large quantity of firecrackers and connected their detonation fuses to a motorcycle battery. The two perpetrators proceeded to start the engine. The resulting explosion could be heard from a distance of two kilometers.

Onlookers attempted to rescue Prasad and Anan, but their burns were too severe. Both men died at the scene. Eight onlookers were treated at a local hospital for their injuries.

DARWIN AWARD: JUSTICE IS SERVED
Confirmed by Darwin
8 APRIL 1999, GEORGIA

William was forty-five years old and boasted a string of arrests and criminal convictions. You would be stretching the truth if you called him an exemplary citizen. But even worse, he was an abusive husband. Charges of battery and false imprisonment had been leveled against him for allegedly tying his common-law wife to their bed with an extension cord in 1997.

According to the police report he told her "he was not going to let her out, feed her, or allow her to go to work." Rosemary managed to free herself and crawl out a window to summon police. William was incarcerated for two months, which he spent writing poems and drawing pictures of the couple in wedding finery, before Rosemary begged a judge to send him home because she couldn't afford the rent alone.

One day in April, William doused Rosemary with gasoline and set her on fire. Then he wrenched a gas line loose, apparently to make her death look like an accident. But his mislaid plan backfired when the gas line ignited and blew him up, putting an end to his boorish behavior.

Reference: Associated Press, News Tribune Company

DARWIN AWARD: NO SMOKING? HA!
Unconfirmed by Darwin
20 MAY 1998, LOUISIANA

A respiratory patient in an oxygen tent at a New Orleans hospital sneaked a pack of cigarettes into his room. One morning in the quiet dawn, the sixty-one-year-old patient ignored the nurses' lectures, ignored the warning signs, and surreptitiously lit his last cigarette.

In the presence of extra oxygen, even a small spark can ignite a flash fire. Before he could even draw a breath of nicotine, the cigarette set his clothes on fire and the flames began to spread. The man, afraid of being caught, tried to extinguish the blaze without sounding an alarm. A hospital employee walked by his room and noticed the man, standing in the midst of a conflagration, quietly trying to pat out the flames.

An orderly carried the patient into the hallway and extinguished his hospital gown with a blanket, while nurses used fire extinguishers to beat back the flames enough to reach the valve and turn off the oxygen supply. Twenty-one patients were evacuated, and seven others were treated for smoke inhalation.

The cause of the blaze was airlifted to the Baton Rouge burn unit with third-degree burns over forty percent of his body, where he died five days later. Was the patient cured of his

> Smoke, smoke, smoke that cigarette . . .
> Tell St. Peter at the Golden gate
> That you hate to make him wait
> But you gotta have another cigarette.
> —Tex Williams lyrics, 1947

addiction by his experience? Apparently not. A pack of Kool cigarettes and a lighter were found hidden in his sock at the burn unit.

Reference: *New Orleans Times-Picayune*

Oxygen is required for combustion. An example tried by many schoolboys is burning methane, present in flammable farts, in the presence of oxygen to yield carbon dioxide and water.

$$CH_4 + 2O_2 \rightarrow CO_2 + 2H_2O + heat$$

The higher the concentration of oxygen, the easier it is for combustion to occur. Our atmosphere is twenty-one percent oxygen, and scientists think that the level has remained fairly constant despite the abundance of oxygen-producing plants, because a higher percentage would cause forest fires to burn more intensely, consuming oxygen in the process.

DARWIN AWARD:
DYNAMITE AND BOATS DON'T MIX
Confirmed by Darwin
16 JUNE 1998, ILLINOIS

A man drowned in Fox Lake after he and a friend inadvertently blasted a hole in the bottom of their rowboat with a quarter stick of dynamite. Daniel, twenty-nine, and his unidentified friend were relaxing on the lake on Sunday in a fourteen-foot aluminum boat, when they decided to toss the M-250 explosive into the water. They intended to kill fish with the blast, not themselves, said chief deputy coroner Jim Wipper. A sudden gust of wind pushed their boat over the firecracker, and the boat sank about a hundred yards from shore. Daniel drowned; the friend swam safely to land.

Reference: Associated Press, *San Francisco Examiner*

DARWIN AWARD: UP IN SMOKE

Confirmed by Darwin

5 MARCH 1999, ENGLAND

Christopher arrived at his Fleet, Hampshire, home with a case of beer. "He drank a quantity of the beer," his wife, Jacqueline, said, "and then started smoking." Was he suicidal or simply stupid?

His drinking binge progressed. At some stage of inebriation Jacqueline observed him clumsily attempting to fill his butane lighter, spilling the flammable liquid on his jumper. She warned him that he was being silly, and she didn't mean amusing. He paid her little heed.

The thirty-five-year-old man flicked his lighter experimentally, then gave in to the lure of pyromania and began trying to burn his trousers. As a side effect he set his fuel-soaked jumper ablaze, turning into a fireball in his own living room!

If you should be so unfortunate as to find yourself ablaze, remember to drop and roll to suffocate the flames.

Christopher did not drop and roll. He flailed in terror and dived from the window into the street, setting fire to curtains and a BMW parked nearby as he attempted to beat out the flames with his hands. His efforts added more oxygen to the combustion, and the flames grew higher.

A neighbor mistook the blaze for a bonfire, but quickly realized that it was a burning man. He rushed from his home and attempted to suffocate the fire with bath towels, to no avail.

Readers complain that this story is unlikely. Butane is a liquefied gas that evaporates rapidly under normal pressure and, while it's not impossible to set yourself on fire with it, it's more likely you would set off an explosion. It is probable that the lighter fluid the victim was using was actually one of the heavier fluids, akin to that which feeds a Zippo flame.

The fuel-fed fire was so hot that it burnt virtually every inch of Christopher's body, all save the soles of his feet. He died shortly after arriving at Frimley Park Hospital in Surrey. The verdict at the Hampshire inquest was accidental death.

Reference: The *Times* of London

DARWIN AWARD: CIGARETTE LIGHTER TRIGGERS FATAL EXPLOSION

Confirmed by Darwin

4 DECEMBER 1996, INDIANA

A Jay County man using a cigarette lighter to check the barrel of a muzzle-loader was killed when the weapon discharged in his face, sheriff's investigators said. Gregory, nineteen, died in his parents' rural Dunkirk home about 11:30 P.M.

Investigators said Gregory was cleaning a .54-caliber muzzle-loader that had not been firing properly. He was using the lighter to look into the barrel when the gunpowder ignited.

Reference: *Indianapolis Star*

DARWIN AWARD: LIGHTS OUT
Confirmed by Darwin
13 AUGUST 1999, CALIFORNIA

On Friday the thirteenth, Scott had an electrifying experience while attempting to view the annual Perseid meteor shower. The aspiring astronomer set up his telescope for a closer view of the sky. Alas, poor Scott did not reflect on the merits of using a telescope for watching the Perseids. A telescope is really a hindrance. The wide field of vision a naked eye apprehends will catch far more shooting stars, particularly if that eye is taken away from city lights into the desert or mountains.

Having already proven to be a poor astronomer, Scott proceeded to show that he was not much of an electrician either. Bothered by the glare of a nearby streetlight, he used pliers to pry open an inspection plate at the base of the light, then sawed into its four-thousand-volt power cord. His sister Kimberly saw a flash knock him onto his back. He was pronounced dead at Hoag Memorial Hospital shortly after his spectacularly aborted sky-watching attempt.

Scott had the technical know-how to construct a computer from scratch or wire a burglar alarm. "He was trying to solve a problem and not using his head, and he made a mistake," grieved the dead man's father. "He didn't realize the power." A friend of Scott's countered, "Scott had an itch for doing things with his hands. He has done many dangerous things. This time he made a fatal mistake."

In the words of a spokesman for Southern California Edison, "This is another example of why you shouldn't tamper with electricity if you don't know what you're doing." There were no shooting stars for Scott that Friday the thirteenth, but he did earn a shot at winning a Darwin Award.

Reference: *Fresno Bee, Los Angeles Times,* CBS News, *Orange County Register,* Channel2000.com, Los Angeles *Press-Telegram*

Read Scott's last conversation, minutes before his death.
Internet Relay Chat Log 13 August 1999
www.DarwinAwards.com/book/lights.html

HONORABLE MENTION: CHIMNEY SAFETY

Confirmed by Darwin

1999, WASHINGTON

A married couple wanting to keep their home fires burning decided to install a woodstove in their Granite Falls home. They figured it didn't take a rocket scientist to install this basic bit of heating hardware, so instead of hiring a professional, they brought the stove home and installed it themselves.

See the Darwin Award "Firefighters Ignite!" on page 107 for more on the history of Snohomish County firefighters.

They even remembered to cut a hole through the ceiling for the chimney vent. Unfortunately they neglected to extend the chimney through the attic to the roof. Pleased with a job well done, they settled down to a cozy evening in front of the fire. And the inevitable happened. The heat and sparks built up in the attic and set their home ablaze, providing an unexpected source of warmth from above.

Snohomish County firefighters extinguished the fire, and the couple returned to their home to console each other over their eight-thousand-dollar loss. But the fire was not quite out. Firefighters had failed to fully extinguish the fire, which started up again the next morning, burning the house to the ground. The husband and wife survived.

Reference: Daily Herald Company, Everett, Washington

HONORABLE MENTION:
I JUST FLICKED MY BIC!
Confirmed by Darwin
28 FEBRUARY 2000, DELAWARE

A Dover man filled his portable propane bottle at a service station, placed the bottle on the floor of the passenger's side, and drove home. As he was driving, the nicotine blues hit him hard. He had to have another cigarette. Unfortunately for him, he had only partially sealed the propane bottle's shutoff valve. Our hero flicked his Bic, and we had liftoff— of the sunroof and windows in his car! Our astronaut didn't make it into outer space this time, but he did manage a ride on the helicopter that airlifted him to the hospital for treatment of his burned hands and face.

Reference: *Dover Post*

URBAN LEGEND:
SCUBA DIVERS AND FOREST FIRES
1998, CALIFORNIA

So you think you're having a bad day?

In California, wildfires are part of the natural cycle of the forest. They are caused by lightning, by arson, by acts of God. Brave firefighters earn their livings extinguishing these ravenous blazes.

Recently, fire marshals found a corpse in a rural section of California while they were assessing the damage done by a recent forest fire. The deceased male was dressed in diving gear consisting of a recently melted wet suit, a dive tank, flippers, and face mask. Apparently the man had been participating in recreational diving fairly recently.

A postmortem examination attributed his death not to burns, but to massive internal injuries. Salt water was found in his stomach. Dental records provided a positive identification of a man who had been reported missing a week before, and the next of kin were notified. Investigators then set about determining how a fully clad diver ended up in the middle of a forest fire.

It was discovered that, on the day of the fire, the deceased had set out on a diving trip in the Pacific Ocean. His third dive was twenty kilometers away from the location of a large brush fire, which was threatening the safety of a nearby town.

Firefighters, seeking to control the conflagration as quickly as possible, had called in a fleet of helicopters to saturate the area with water. The helicopters towed large

buckets, which were dropped into the ocean for rapid fill-
ing, then flown to the fire and emptied.

You guessed it! One minute our diver was marveling at
the fish species of the Pacific, and in the next breath he
found himself in a fire bucket three hundred meters in the
air. He experienced rapid decompression caused by the altitude change, suddenly followed by a plummet into burning
trees.

As a consolation to bereaved relatives, investigators calcu-
late that the man extinguished 1.78 square meters of the fire,
approximately the area covered by a splattered human body.
Bereaved are also consoled by the knowledge that he had enjoyed two rewarding dives preceding his fatal third dive.

Divers and pilots alike are being warned to remain on the
alert. Divers are encouraged to remain calm if scooped from
the water, and to hang on to the bucket when the water is
dumped on the fire. Decompression chambers will be made
available immediately upon landing.

URBAN LEGEND: COW BOMB
1999, CALIFORNIA

A dairy worker who heard that bovine flatulence was largely composed of methane, and potentially explosive, decided to apply the scientific method to the theory. While one of his contented cow charges was hooked up to the milking machine, he waited for the slight tail lift that dairy workers know signals an impending expulsion, generally something one avoids. Our hero struck a match.

His satisfaction at seeing the resulting foot-long blue flame lasted mere seconds, before the flame was subsumed by a rectal contraction. The poor Holstein exploded, killing the worker, who was struck by a flying femur bone.

This story is just a legend. Since methane combusts according to the equation, $CH_4 + 2O_2 \rightarrow CO_2 + 2H_2O$, there would have to be twice as much oxygen as methane inside the cow for it to combust—unlikely, since digestive processes do not create oxygen. Hospitals do occasionally see people who have rectal burns from this kind of shenanigan, but lit farts simply do not produce the force needed to explode an organic body. Besides, if this story were true, imagine what would happen if the dairy farmers became angry with the government. News headlines would read, "Internal Revenue Service Destroyed by Cow Bomb."

Urban Legend: Raccoon Rocket
1997, Pennsylvania

A group of men were drinking beer in rural Carbon County, and discharging firearms from the rear deck of a home owned by twenty-seven-year-old Leon. The men were firing at a raccoon that had the misfortune to wander by, but the beer must have impaired their aim. Despite thirty-five shots launched by the group, the animal escaped into a three-foot-diameter drainage pipe a hundred feet away from his deck.

Determined to terminate the animal, Leon retrieved a can of gasoline and poured some down the pipe, intending to smoke the animal out. After several unsuccessful attempts to ignite the fuel, Leon emptied the entire five-gallon fuel can down the pipe and dropped in another match, to no avail.

Not one to admit defeat at the hands of wildlife, the determined Leon proceeded to slide feet-first approximately fifteen feet down the sloping pipe to toss the match. This time he was successful. The subsequent rapidly expanding fireball propelled Leon back the way he had come, though at a much higher rate of speed. He exited the angled pipe "like a Polaris missile leaving a submarine," according to one witness.

Leon was launched directly over his own home, right over the heads of his astonished friends, onto his front lawn. In all he traveled over two hundred feet through the air.

"There was a Doppler effect to his scream as he flew over us," a witness reported, "followed by a loud thud." Amazingly, Leon suffered only minor injuries.

"It was actually pretty cool," the human cannonball said. "Like when they shoot someone out of a cannon at the circus. I'd do it again if I was sure I wouldn't get hurt."

URBAN LEGEND:
HYDROGEN BEER DISASTER
1999, TOKYO

The recent craze for hydrogen beer is at the heart of a three-way lawsuit between unemployed stockbroker Toshira Otoma, the Tike-Take karaoke bar, and the Asaka Beer Corporation. Mr. Otoma is suing the bar and the brewery for selling toxic substances, and is claiming damages for grievous bodily harm leading to the loss of his job. The bar is countersuing for defamation and loss of customers.

The Asaka Beer corporation brews Suiso brand beer, in which the carbon dioxide normally used to add fizz has been replaced by the more environmentally friendly hydrogen gas. Two serendipitous side effects of the hydrogen gas have made the beer extremely popular at karaoke sing-along bars and discotheques.

First, because hydrogen molecules are lighter than air, sound waves are transmitted more rapidly, so individuals whose lungs are filled with the nontoxic gas can speak with an uncharacteristically high voice. Exploiting this quirk of physics, chic urbanites can now sing soprano parts on karaoke sing-along machines after consuming a big gulp of Suiso beer.

Second, the flammable nature of hydrogen has also become a selling point, though it should be noted that Asaka has not acknowledged that this was a deliberate marketing ploy.

The beer has inspired a new fashion of blowing flames from one's mouth using a cigarette as an ignition source. Many new karaoke videos feature singers shooting blue flames in slow motion, while flame contests take place in pubs everywhere.

"Mr. Otoma has no one to blame but himself. If he had not become drunk and disorderly, none of this would have happened," said Mr. Takashi Nomura, manager of the Tike-Take bar. "Mr. Otoma drank fifteen bottles of hydrogen beer in order to maximize the size of the flames he could belch during the contest. He catapulted balls of fire across the room that Godzilla would be proud of, but this was not enough to win him first prize since the judgment is made on the quality of the flames and the singing, and after fifteen bottles of lager he was badly out of tune.

"He took exception to the contest results and hurled blue fireballs at the judges, scorching a female judge's hair and entirely removing her eyebrows and lashes, and ruining the clothes of two nearby customers. None of these people have returned to my bar. When our security staff approached Mr. Otoma, he turned his attentions to them, making it almost impossible to approach him. Our head bouncer had no choice but to hurl himself at Mr. Otoma's knees, knocking his legs from under him.

"The laws of physics are not to be disobeyed, and the force that propelled Mr. Otoma's legs backward also pivoted around his center of gravity and moved his upper body forward with equal velocity. It was his own fault that he had his mouth open for the next belch, his own fault that he held a lighted cigarette in front of it, and his own fault that he swallowed that cigarette.

"The Tike-Take bar takes no responsibility for the subsequent internal combustion and rupture of his stomach lining, nor the third degree burns to his esophagus, larynx, and sinuses as the exploding gases forced their way out of his body. Mr. Otoma's consequential muteness and loss of employment are his own fault."

Mr. Otoma was unavailable for comment.

URBAN LEGEND:
CELL PHONE DESTROYS GAS STATION
1999

Finally a solution to the obnoxious cell phone driver problem!

A motorist suffered severe burns recently when his cell phone ignited gasoline fumes at a gas station. "Read your handbook!" manufacturers admonish. Motorola, Ericsson, and Nokia all print cautions in their user handbooks, warning against the use of mobile phones near gas stations, fuel storage sites, chemical refineries, and nuclear reactors.

Electronic devices in use at gas stations are protected with explosive containment devices, making them safe to use around volatile hydrocarbons. Cell phones and other high-voltage battery appliances, on the other hand, are not shielded. They are in clear danger of producing small sparks.

Exxon has begun to place warning stickers on its gasoline pumps.

Readers note that the cell phone explosion myth was thoroughly investigated and debunked on National Public Radio in November 1999. Oil companies are reportedly still pressing ahead with their warning sticker campaigns. A safety consultant who was privately commissioned to study the danger concluded, "There is no evidence that this has happened, could happen, or is even physically possible. The probability of a gaseous explosion ignition from a cell phone is 4.01×10^{-18}." It is pure Urban Legend.

Will your cell phone explode? Know the facts.
www.DarwinAwards.com/book/cellphone.html

PERSONAL ACCOUNT: ELEMENTAL MISTAKE
13 JULY 1999

Place a small piece of elemental sodium in a tank of water. The reaction is so violent that a small fire erupts. In fact, sodium is stored submerged in oil, to prevent a fire from being sparked by water vapor. High school chemistry teachers often perform this riveting demonstration without realizing that a pyromaniac lurks in their classrooms.

> Sodium and water react to form flammable hydrogen gas, hence the fire, and sodium hydroxide, which is a powerful base and causes severe caustic burns to the skin.

One chemistry student was particularly taken with this demonstration, and contrived to acquire a piece of elemental sodium for his own use. He extracted a chunk from the canister of oil, wrapped it in a paper towel, and hid it in his pants pocket. He clearly intended to play with the contents of his pocket, later, in private.

But as soon as the oil drained off, the sodium began to react with all nearby sources of water. Humans are sixty to seventy percent water, so in this case the nearest source of water was the student's leg, which was badly burned by the ensuing conflagration.

Reference: Reece R. Pollack, personal account.

PERSONAL ACCOUNT: IS IT LOADED?
FEBRUARY 2000

Years ago I was visiting my local gun range when I noticed a gentleman having trouble with his muzzle-loader.

You load a muzzle-loader by pouring black gunpowder down the barrel, placing a bullet on top of the powder, and ramming it down tight with a rod. You shoot the loaded gun by cocking the hammer and placing a pinch of gunpowder under it. When the trigger is pulled, the hammer falls against a flint. The resulting sparks ignite the powder, a jet of flame flashes through a hole into the main charge, and the gun goes bang.

This gentleman loaded his gun, but when he pulled the trigger, nothing happened. He tried it again, and again nothing happened. So he looked down the barrel. It was dark in there and he could not see anything. So he pulled out his lighter and held the flame against the flash hole (where the pinch of powder is placed) while *still* looking down the barrel.

I ran up and separated idiot from gun before anything. could happen. No waiting-period gun law could save this fellow from himself.

The cause of the malfunction turned out to be old, degraded powder.

Reference: Albert D. Mayse, personal account.

PERSONAL ACCOUNT: FINAL FLICK OF BIC
1970S, INDONESIA

A missionary to Indonesia two decades ago was not just a preacher, but a teacher as well. Indonesia, like many former colonies, had nationalized their petroleum industry to keep profits closer to home. Local workers were hired to run the plants, but most equipment was labeled in English because the previous supervisors were from the United States. The plant hired the missionaries to tutor new workers in English.

One day there was a commotion at the gate and only a handful of students showed up for class. They were obviously shaken, jabbering about a mandatory safety demonstration and a "big showing" by the safety inspector.

"Big showing" was putting it mildly.

After lecturing for an hour the safety inspector had taken twenty workers out into an open field to demonstrate what *not* to do around an oxygen cylinder. He admonished his students to stand way back, and he would show them how something they couldn't see or smell could hurt them.

The instructor took a Bic lighter from his pocket, opened the petcock on the oxygen tank, and in a final dramatic gesture, he flicked his Bic. It was a most convincing safety demonstration The largest piece of the safety engineer found afterward was the size of a postage stamp.

The students translated the mishap into a pidgin English: "The tank was boom and inspector was not so good after that!"

Reference: Anonymous personal account.

Scientific Merits: Oxygen is stored compressed in tanks, not lique-fied, at 2,200 psi. It is unlikely that an oxygen tank will explode if an open flame is used near the valve. Oxygen does not burn, it merely promotes rapid oxidation, and is termed an "accelerant." Holding a lit Bic to the open valve would at best make a larger flame. It is more likely that the oxygen would simply blow out the flame, since the stream of gas is traveling at close to the speed of sound. These points show discrepancies between the story and the realm of science. However, it is possible that the tank contained a flammable gas rather than oxygen.

CHAPTER 5

Leaps of Faith:
Fatal Falls

All men can fly, but sadly, only in one direction.

JOHN F. KENNEDY JR.

Humans seem to possess an innate love of flying, and yet time and again our bodies prove woefully inadequate to the job. These stories show what happens when a lamentable miscalculation, combined with a lack of respect for gravity, leads to its inevitable conclusion: man is not meant to fly.

One of the most controversial discussions in the history of the Darwin Awards followed the death of a well-known American. John F. Kennedy Jr. crashed his plane into the sea near Martha's Vineyard on July 16, 1999, killing himself and two passengers. A spontaneous debate on the merits of honoring him with a Darwin Award quickly emerged on the Philosophy Forum.

The Philosophy Forum is a focal point of the Darwin Awards culture. Readers gather to express their opinion of nominees, discuss the role of natural selection in human biology, and explore ethical questions arising from the theory of evolution. The heated debate following JFK Jr.'s crash is an entertaining example of this community in action, and illustrates the nominee selection process.

Condensed from the National Transportation and Safety Board Preliminary Report NYC99MA178.

On July 16, 1999, a Piper was destroyed during a collision with water near Martha's Vineyard, Massachusetts. The pilot and two passengers were fatally injured. Night visual conditions prevailed, and no flight plan had been filed for the personal flight.

A person using the pilot's log-in code obtained aviation weather information from an Internet site three hours before the crash. The forecast called for winds at ten knots, visibility six miles, and clear sky. No AIRMETS or SIGMETS were issued for the route of flight, and all airports along the route reported visual meteorological conditions.

The pilot received his private pilot certificate in April 1998. He did not possess an instrument rating. Interviews and training records revealed that the pilot had accumulated about three hundred hours of total flight experience, not including recent experience gained in the accident airplane.

On July 20, 1999, the airplane was located in 116 feet of water. Preliminary examination of the wreckage revealed no evidence of in-flight structural failure or fire, nor of conditions that would have prevented either the engine or propeller from operating. Pilots who flew over Long Island Sound that evening were interviewed after the accident. They reported that the in-flight visibility over the water was significantly reduced.

Full Text of the NTSB Preliminary Report:
www.DarwinAwards.com/book/kennedy.html

CHARRELL6170

After due consideration I have decided to nominate JFK Jr. for a Darwin Award. Here is a man who held his private pilot's license for only fourteen months, and was not cleared for instrument flight. Nevertheless, he was flying at night in a high performance aircraft with which he was unfamiliar. He flew with no flight plan, not illegal but ill advised, under reduced-visibility conditions. Based on these merits I feel that he has earned a Darwin Award.

PSYCHOTIK

We wouldn't nominate him if it were a car accident, why a plane accident? There isn't anything Darwinian about the way he snuffed himself, and it's hardly entertaining. The Darwin Award winner shows more than simple bad decision making. This one doesn't get my vote.

HUGO

If JFK Jr. is nominated, then I retroactively nominate Amelia Earhart for a Historical Darwin Award.

UK YANK

Of course he should be included! Let me put this in context for you nonpilots. It's the night before your relative gets married and you need to get to the wedding. Your only vehicle is a blindingly fast Porsche 911 Turbo, an accident waiting to happen in the hands of a new driver such as yourself. You have no snow chains, a foot that's not fully healed, and it's getting dark. Your driving buddies tell you thick fog is expected, and enough snow to hide the road. They wouldn't dream of going out on a night like this. What do you do?

a) Hire a qualified racing driver to drive your Porsche.
b) Wait for morning and better driving conditions.
c) Jump in the Porsche and hope for the best.

Gimme a break. When you know the odds are against you, yet still risk your life, you're not drinking from the fountain of wisdom—you're just gargling. Give the man a prize!

TREVORG

I investigated aviation accidents for the military, and this case hardly needs investigating. The answer is obvious. He flew right off the page that details "predictable ways for stupid pilots to die." To commit such a grievous error in the face of such a mountain of information makes JFK Jr. a solid contender for the Darwin Award.

SREDDY17

He was not acting stupid. The control tower knew how much experience he had. If they thought he was incapable of flying, they wouldn't have let him out of the airport. Besides, if he had the least apprehension about his abilities, he wouldn't have placed his wife and sister-in-law in danger.

SIDECAR

Those with flying experience know that the pilot is master of his own destiny. When he warms up his plane and taxis to the edge of the runway, the control tower gives him information including wind speed and direction. When it is clear for him to take off, the tower uses the following words: "You are cleared to take off at your discretion." The words *at your discretion* absolve the tower of responsibility for the pilot or aircraft. All contact with air control agencies is for information only. It's up to you to act on this information responsibly.

SReddy17

The guy's dead, show a little respect. Maybe something totally out of control and strange happened, who knows? Even pilots with years of experience have been known to crash on nights exactly like this one, so cut a little slack.

A Voice in the Wilderness

The evidence is against you, SReddy17. You stated the reason yourself in your defense of JFK Jr. "Pilots with years of experience have been known to crash on nights exactly like this one." That is exactly why he should not have flown the aircraft.

Darwin

The controversy over the pros and cons of whether to nominate John F. Kennedy Jr. for a Darwin Award makes it a difficult decision. I have weighed the comments, ignored the media hype, and decided to disqualify this nomination. "There are old pilots and there are bold pilots, but there are no old bold pilots." The common adage illustrates that his was a common lapse of judgment, therefore JFK Jr.'s actions are not worthy of a Darwin Award.

DARWIN AWARD: DON'T DRINK AND FLY
Confirmed by Darwin
25 APRIL 1998, MASSACHUSETTS

One fateful day in April a private pilot landed his Piper plane at the New Bedford Airport. To secure his aircraft against thieves, he inserted a gust lock into the copilot control column and padlocked it in place. This procedure is fairly common, except that the gust lock is usually placed on the pilot control column. That way it's hard to forget to remove it when you prepare to depart. Many gust locks have a big red plate that hangs down to cover the ignition and master switch. We will never know why our soon-to-be-dead friend chose to put the gust lock on the copilot's side.

The pilot went off to have some drinks and returned, somewhat potted, to his plane at 10:30 P.M. He hopped into the aircraft with 155 mg/dl of ethanol in his blood, nearly four times the legal limit, and departed without remembering to check that the flight controls were unobstructed. A witness to the accident reported that he departed the runway at a very steep angle, consistent with having a gust lock installed.

At about this time our erstwhile friend realized that he had forgotten to remove the gust lock, and that his plane would soon stall. The problem was that the key for the padlock was on the same key ring as the key for the ignition. So he had two choices: try to remove the padlock key from the key ring while keeping the plane running, which would take more time than he had, or turn off the engine, which would accelerate the stall, then rush to remove the gust lock and restart the engine. He chose the second option.

But he didn't restart the engine in time. The airplane, its course fixed by the gust lock, "went straight up in the air like an acrobat" then appeared to level off, turned northwest, then northeast, and completed its ballet with a nosedive and a rapid descent to the ground.

When the National Transportation Safety Board investigator arrived at the scene he discovered the padlock and gust lock still installed, and the key ring with both keys still on it on the floor of the cockpit.

Reference: National Transportation Safety Board

Read the full accident report.
www.DarwinAwards.com/book/fly.html

DARWIN AWARD: LAWYER ALOFT
Confirmed by Darwin
1996, TORONTO

Police said a lawyer demonstrating the safety of windows in a downtown Toronto skyscraper crashed through a pane of glass with his shoulder and plunged twenty-four floors to his death. A police spokesman said Garry, thirty-nine, fell into the courtyard of the Toronto Dominion Bank Tower as he was explaining the strength of the building's windows to visiting law students. Garry had previously conducted the demonstration of window strength without mishap, according to police reports. The managing partner of the law firm that employed the deceased told the *Toronto Sun* newspaper that Garry was "one of the best and brightest" members of the two-hundred-man association.

Reference: UPI

RealAudio presentation of Lawyer Aloft
www.DarwinAwards.com/book/realaudio2.html

DARWIN AWARD: STONED SLEEP
Confirmed by Darwin
26 MARCH 2000, SOUTH CAROLINA

A North Carolina woman learned a hard lesson about drugs when she decided to sleep on the roof. Police reports say that Patricia and her boyfriend had been drinking and smoking marijuana, when they decided to enjoy the fresh air on the roof of the King Charles Inn. They climbed over a guardrail with pillows and blankets, and fell asleep under the stars. Sound asleep, apparently. Patricia slid off the roof and fell to her death on Hasell Street shortly before dawn on Sunday. When police arrived at the scene, the boyfriend was found still sleeping on the roof, curled up in a blanket and pillow. The death has been ruled accidental, but we feel that the blame lies with the stoned woman who chose to snooze on the roof.

References: News and Observer Publishing Company,
Associated Press, Charleston.net

DARWIN AWARD: HOMEGROWN PARACHUTE
Confirmed by Darwin
25 MAY 2000, PHILIPPINES

We all enjoy learning from the past. Reflect back to November 24, 1971, when a gentleman in a dark suit carrying a briefcase boarded a Northwest Orient Airlines flight in Portland. He reclined in seat 18F and passed a note to the flight attendant, asking her to sit next to him because he had a bomb. The attendant, Flo Schaffner, thought the man was giving her his telephone number, and stuck the note in her apron. It wasn't until later that Flo actually read the hijack note.

Once alerted to the irregularity, she and the other attendants relayed notes from the man, who had purchased his ticket under the name of "Dan Cooper," to the cockpit. Cooper demanded two hundred thousand dollars in cash and four parachutes. The plane made a landing at the Seattle-Tacoma Airport to accommodate his requests, and most of the flight attendants and passengers were allowed to disembark from the plane. One flight attendant and the pilots remained.

Cooper asked to be flown to Mexico, and agreed to stop for fuel in Reno. While en route to Reno, he asked the flight attendant how to lower the tail stairs on the Boeing 727. He then requested her to join the others in the forward cabin and pull the curtains shut.

When the plane landed in Reno, the tail stairs were open and Cooper and the money were gone. He was mistakenly named D. B. Cooper by an FBI agent, and the legend of D. B. Cooper survives to this day. Cooper has never been found, but in 1980 a boy playing in a creek in Washington found $5800 of his cash.

For all his cool demeanor, Cooper had the crosshairs of evolution on him when he decided to jump. There was a freezing rainstorm outside, and the wind chill from the plane's velocity dropped the effective temperature to minus sixty degrees Fahrenheit. To seal his fate, he jumped with no food or survival gear into a heavily wooded forest in winter at night.

The peanuts provided on the plane were just not enough to sustain his life. It is assumed that he died in the mountains or hit the Columbia River and drowned. History, then, teaches us that one cannot jump out of an airplane and survive. You would think that a hijacker would know better, but . . .

We turn to Davao City in the Philippines on May 25, 2000. Augusto was a man with a mission. He boarded a Philippine Air flight to Manila, and donned a ski mask and swim goggles. Then he pulled out a gun and a grenade and announced that he was hijacking the plane. Apparently security is a bit lax at the Davao City airport.

He demanded that the plane return to Davao City, but the pilots convinced him that the plane was low on fuel, and they continued on toward Manila. Augusto, undaunted, robbed the passengers of about $25,000 and ordered the pilots to lower the plane to 6,500 feet.

When a lunatic with a gun orders you to descend, you descend. Meanwhile, Augusto strapped a homemade parachute onto his back, and forced the flight attendants to open the door and depressurize the plane.

He probably intended to jump, but the wind was so strong that he had trouble getting out of the plane. Finally one of the flight attendants helpfully pushed him out the door, just as he pulled the pin from the grenade. He threw the pin (oops!) into the cabin, and fell toward the earth carrying the business end of the grenade in his hand.

The impact of Augusto hitting the earth at terminal velocity had little impact on the earth's orbit. The ground embraced him with an enthusiasm that said, "Hey, we won't have to dig a burial plot for you!" All that remained above-ground were Augusto's two hands.

So history repeats itself with a new twist to the lesson:

Lesson 1: Don't throw yourself out of a perfectly good airplane.

Lesson 2: If you feel compelled to violate Lesson 1, at least don't roll your own . . . parachute, that is.

Reference: Associated Press, *Australia Age*, Reuters, *National Enquirer*

DARWIN AWARD:
YOSEMITE PARACHUTE SAFETY
Confirmed by Darwin
22 OCTOBER 1999, CALIFORNIA

Easy as falling off a cliff.
Yosemite National Park bans parachuting from its majestic
cliffs, citing the dangers inherent in the practice. But those
cliffs are too challenging for BASE (building, antenna, span,
earth) jumpers to ignore. Every year men and women surreptitiously prepare for daredevil plunges into the abyss.
Every year park rangers hunt them down, confiscate their
gear, arrest them, and fine them two thousand dollars.

On this fateful day in evolutionary history, activists hoping to persuade officials to open Yosemite for parachuting
had arranged a demonstration to showcase the safety of
BASE jumping. A group of five climbed 3,200-foot El Capitan Peak, while hundreds of people watched from below.

One by one the parachutists jumped over the edge,
pulled their rip cords, and floated safely to the valley floor.
Until our candidate, an experienced skydiver with a borrowed parachute, gave an astounding demonstration of the
dangers inherent in skydiving. She leapt off the cliff and
plummeted directly to the ground without deploying her
parachute. Her death was captured on film by her traumatized husband.

What went wrong?

Adventure Athletes had arranged the jumps with grudging

cooperation from park officials, who were concerned for the safety of visitors in the vicinity of the drop zone. The protesters agreed to be arrested and fined after the jump, and have their equipment confiscated.

The deceased, a skydiver prominent in the extreme-sports community, had numerous jumps under her belt. She was loath to part with her valuable equipment, so she borrowed an inexpensive parachute for the jump. This chute had a rip cord on the leg, rather than the back, but because she didn't give the parachute a basic safety check, she was unable to find the cord in midair. Placing financial considerations above safety concerns may have cost her her life when she landed on the gigantic pile of talus at El Capitan's base.

BASE jumping is technically difficult because the jump-off point is close to the ground and it is performed in tight spaces. Six jumpers have died BASE jumping in Yosemite, including a parachutist who drowned in the Merced River while eluding park rangers.

Her husband reportedly vowed to continue the protest against Yosemite's BASE-jumping ban, thus qualifying him for an Honorable Mention.

Reference: *San Jose Mercury News*, Associated Press
San Francisco Examiner, New York Times

DARWIN AWARD: SHOCKING FALL
Confirmed by Darwin
1 JANUARY 2000, NEVADA

Tod made a place for himself in history by being the first person to die in Las Vegas while celebrating the new millennium. Minutes before midnight the twenty-six-year-old Stanford graduate climbed to the top of a streetlight in front of the Paris Las Vegas Hotel and waved to the enthusiastic revelers below. At the stroke of midnight he slipped and, in an effort to break his fall, grabbed the wires that were supplying electricity to the streetlight. Suddenly he was conducting more than a cheering crowd.

A news camera caught his foolhardy climb and subsequent headfirst plunge to the concrete below. It has not yet been determined whether he died from electrocution or from the thirty-foot fall, but either way he earns the first Darwin Award of the new millennium.

Clearly, a sterling academic pedigree is no indication of common sense. Tod was a Stanford graduate working at a Silicon Valley start-up. Before his departure for Vegas, a friend warned, "People are going to be doing crazy things. Be careful."

Tod replied, "You know I won't."

Reference: *Las Vegas Sun, Yahoo! News*, KCBS,
Associated Press, *San Jose Mercury News*

DARWIN AWARD: BUNGEE JUMPER
1997 Darwin Award Winner
Confirmed by Darwin
13 JULY 1997, VIRGINIA

Eric, a twenty-two-year-old Reston resident, was found dead after he used Bungee cords to jump off a seventy-foot railroad trestle, police said. The fast-food worker taped a number of Bungee cords together and strapped one end around his foot. Eric remembered to measure the length of the cords to make sure that they were a few feet short of the seventy-foot drop, and he had the foresight to anchor the bitter end to the trestle at Lake Accotink Park. He proceeded to fall headfirst from the trestle, and hit the pavement seventy feet below several seconds later.

Fairfax County police said, "The stretched length of the cord that he had assembled was greater than the distance between the trestle and the ground."

Perhaps the deceased fast-food worker should have stuck to the line, "Do you want fries with that?"

Reference: *Washington Post*

DARWIN AWARD: BRIDGE BONZAI
Unconfirmed by Darwin
SEPTEMBER 1998, ARIZONA

On a clear Tuesday evening in Phoenix, an intoxicated driver and his passenger were on their way home after a night of partying when their car sideswiped another vehicle. The two drivers pulled off the freeway. At this point one expects to exchange insurance information or contact authorities. Instead, the scene took a melodramatic turn.

The culpable driver turned to his passenger and shouted, "Let's run!"

In shock from the collision, the passenger was unable to budge from the Volkswagen Jetta. The driver, however, was off and running along the freeway shoulder, weaving in a drunken stupor. He reached a nearby overpass, flung his legs over the side of the bridge, and leapt forty-eight feet to his death in the dry Salt River bed below. No one else was injured.

Reference: *Arizona Republic*

DARWIN AWARD: ONE FOR THE BIRDS
Confirmed by Darwin
1996, CANADA

A man cleaning a bird feeder on the balcony of his condominium apartment in the Toronto suburb of Mississauga slipped and fell twenty-three stories to his death, said police. Stefan, fifty-five, was standing on a wheeled chair when the accident occurred, said Inspector D'Arcy Honer of the Peel regional police. Chairs with wheels are notoriously unstable as footstools. "It appears the chair moved and he went over the balcony," Honer said. "It's one of those freak accidents. No fowl play is suspected."

Reference: Reuters

Personal Account: Leap of Faith
March 1966, Georgia

Parachute trainees at the Jump School at Fort Benning were repeatedly warned that while they were floating down under silk, they were to avoid at all costs crossing over another jumper's parachute. Such stunts can kill the unwary.

One day a group of jump trainees stood watching the rest of their platoon make the "leap of faith." And wouldn't you know it, right in front of their very eyes one jumper allowed his chute to drift over another jumper's parachute. It was like a training video in real time!

The bottom parachute stole the air from the top parachute, which collapsed and set the top jumper down on the bottom jumper's canopy. While the top jumper sat briefly on this Jell-O-like pedestal, his own parachute dropped down and dangled around the bottom jumper. Then the still-inflated bottom parachute deformed, and spit the top jumper off into space.

The falling jumper, in fear for his life, grabbed the risers of the bottom parachute, and slid down them, where he ended up face-to-face with the bottom jumper, who proceeded to pummel the intruder in the face. The other jumper began fighting back, and the brawl continued until they hit the ground in a heap.

A member of the United States Parachute Association cautions, "An inflated canopy will not support the weight of a human being. Pictures of people standing on a chute are done under a fully inflated canopy, and it is the top parachute which supports the weight, not the parachute on which they are standing."

The event taught the observers a lesson in parachuting, and the lesson that potential tragedy can be spectacularly funny.

Reference: Mike Fewell, personal account from his third week of Jump School at Fort Benning.

PERSONAL ACCOUNT:
ONE, TWO, THREE, HEAVE!
JANUARY 2000, TEXAS

I work in the petrochemical industry, one of the most hazardous professions on the planet. As a result any aspiring professional in the industry places safely first and foremost. The philosophy of safety was drilled into my head as a summer intern, so I was aware that safety was a mandatory buzzword during subsequent job interviews.

Union Carbide in Sea Drift, Texas, interviewed me for a job in their facility, and I had the opportunity to eat lunch with the plant manager. To impress him with my concern for safety I asked him about their safety program. He said that they had had only one fatality in the history of their plant, and related this story.

In the 1980s a major cleanup had been ordered for the plant. Two people were assigned to remove debris from the top of an oil storage tank, which was large, round, and thirty feet tall. Being safety conscious, the workers were careful to wear harnesses and tie themselves securely before they began removing the debris. Removal involved picking up the junk and dropping it to the ground below.

The team completed most of their task, and was left with one final, particularly heavy piece of debris. It was saved for last precisely because of its large mass. This was definitely a two-man job. They picked up the piece, and agreed to pitch it over the edge on the count of three. The first words heard

after the word *three*, and the resulting heave, were something along the lines of "Oh shit! I'm tied to it."

Reference: Anonymous personal account.

URBAN LEGEND:
MISADVENTURE AT THE METALLICA CONCERT
1996, WASHINGTON

Police in George, Washington, issued a report on the events leading up to the deaths of Robert Uhlenake, twenty-four, and his friend, Ormond D. Young, twenty-seven, at a Friday night Metallica concert.

Uhlenake and Young were found dead at the Gorge Amphitheater soon after the show. Uhlenake was in pickup that was on top of Young at the bottom of a twenty-foot drop. Young was found with severe lacerations, numerous fractures, contusions, and a branch in his anal cavity. He also had been stabbed, and his pants were in a tree above him some fifteen feet off the ground, adding to the mystery of the heretofore unexplained scene.

According to Commissioner-in-Charge Inoye Appleton, Uhlenake and Young had tried to get tickets for the sold-out concert. When they were unable to obtain tickets, the two decided to stay in the parking lot and drink. Once the show began, and after the two had consumed eighteen beers between them, they hit upon the idea of scaling the seven-foot wooden security fence around the perimeter of the site and sneaking in.

They moved the truck up to the edge of the fence and decided that Young would go over first and assist Uhlenake. They did not count on the fact that, while it was a seven-

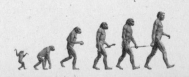

foot fence on the parking lot side, there was a twenty-three-foot drop on the other side.

Young, who weighed 255 pounds and was quite inebriated, jumped up and over the fence and promptly fell about half the distance before a large tree branch broke his fall and his left forearm. He also managed to get his shorts caught on the branch. Since he was now in great pain and had no way to extricate himself and his shorts from the tree, he decided to cut his shorts off and fall to the bushes below.

As soon as he cut the last bit of fabric holding him on the branch, he suddenly plummeted the rest of the way down, losing his grip on the knife. The bushes he had depended on to break his fall were actually holly bushes, and landing in them caused a massive number of small cuts. He also had the misfortune to land squarely on a holly bush branch, effectively impaling himself. The knife, which he had accidentally released fifteen feet up, now landed and stabbed him in his left thigh. He was in tremendous pain.

Enter his friend Robert Uhlenake.

Uhlenake had observed the series of tumbles and realized that Young was in trouble. He hit upon the idea of lowering a rope to his friend and pulling him up and over the

> Urban Legend Status conferred December 31, 1997. Intensive searching of online Washington State newspapers failed to produce validation. The statement attributed to the commissioner is obviously bogus, as police do not make light of deadly shenanigans and never use the word *ass* to describe the rectum. And the Washington State sheriff's office disclaimed knowledge of this story.

fence. This was complicated by the fact that Uhlenake was outweighed by his friend by a good hundred pounds. Happily, despite his drunken state, he realized he could use their truck to pull Young out. Unhappily, because of his drunken state, Uhlenake put the truck in reverse gear. He broke through the fence and landed on Young, killing him. Uhlenake was thrown from the truck and subsequently died of internal injuries.

"So that's how a dead 255-pound man with no pants on, with a truck on top of him and a stick up his ass, came to be," said Commissioner Appleton.

URBAN LEGEND:
POWER PLANT FITNESS FREAK
FEBRUARY 2000

This tale was unveiled during a safety seminar at a power plant in the southern United States. It was a coal-burning power plant, and an employee named Jack had the responsibility of supervising the coal runner. The runner resembled a small treadmill, and transported coal from the hopper to the burner. Jack was stationed near the hopper chute, and watched to make sure nothing blocked the flow of coal, and nothing inappropriate was burned.

One day Jack's coworkers returned from their break to find Jack missing. All that remained was his lunch pail and, curiously, his work boots. No one could explain his continuing absence. After several days the company launched an investigation. The truth came to light, though it took a bit of persuasion to extract the story from his reluctant coworkers.

Jack's doctor had recently warned him that his cholesterol and blood pressure were both dangerously high. The doctor suggested regular mild exercise. Jack had little spare time on his hands, but thought that he could fit in some exercise during his lunch break. He would eat his lunch, then change into sneakers and hop onto the coal runner to jog until his break was over. Because he was self-conscious about his weight, he always made sure nobody was around when he exercised.

Jack's body was never found. Fortunately he had confided his novel exercise regime to a few people at the power plant, or we would never have learned of his tragic demise. Jack must have fallen and been converted into power for hundreds of homes, paving the way for a new, ecologically sound replacement for fossil fuels: Darwin Award contenders.

PERSONAL ACCOUNT: LEVELED
1930S OHIO

An article in the *Cleveland Plain Dealer* describes a most astounding death. A construction worker was busy building a building, when he realized that he needed a three-hundred-pound block suspended above him. But he was the only man on his level, and it normally takes two men to prevent the block from landing too hard on the metal girder. This worker figured it would save time if he just cut the rope and let the block fall forty feet.

The block crashed right through the level he was standing on and killed another fellow below. He leaned over to see if the fellow that the block landed on was okay, and lost his balance. He fell forty feet to the next level, breaking his neck. And so, there was a dead guy underneath the block, and a dead guy on top of it.

Reference: Anonymous personal account and the *Cleveland Plain Dealer*.

CHAPTER 6

Military Intelligence: Uninformed Men

*"The glories of our blood and state
are shadows, not substantial things;
there is no armour against fate.
Death lays his icy hands on kings."*
—James Shirley

HISTORICAL DARWIN AWARDS

The stories in this chapter are about the unique misadventures of police and military men. The Armed Forces have a set of traditional tales that are passed on through generations of enlisted men. These tales evoke the subject of historical Darwins. There have been some famous winners over the past two thousand years:

- Attila the Hun was one of the most notorious villains in history. He conquered all of Asia by 450 A.D. by destroying villages and pillaging the countryside. This bloodthirsty man died from a nosebleed on his wedding night. After feasting and toasting his own good fortune, he was too drunk to notice his nose, and he drowned in a snoutful of his own blood.
- Tycho Brahe, a sixteenth-century Danish astronomer whose research helped Sir Isaac Newton

devise the theory of gravity, died because he didn't make it to the bathroom in time. In those days it was considered an insult to leave the table before the banquet was over. Brahe forgot to relieve himself before the banquet began, then exacerbated matters by imbibing too much alcohol at dinner. Too polite to ask to be excused, he instead allowed his bladder to burst, which killed him slowly and painfully over the next eleven days.

- Francis Bacon was an influential statesman, philosopher, writer, and scientist in the sixteenth century. He died while stuffing snow into a chicken. He had been struck by the notion that snow instead of salt might be used to preserve meat. To test his theory he stood outside in the snow and attempted to stuff the bird. The chicken didn't freeze, but Bacon did, prompting the question "Which froze first? The Bacon or the egg?"

- Jean-Baptiste Lully, a seventeenth-century composer who wrote music for the king of France, died from an overdose of "musical enthusiasm." While rehearsing for a concert, he became overexcited and drove his staff right through his foot. He succumbed to blood poisoning.

Some treasured Historical Darwins are not true. For instance, the legendary circumstances surrounding the death of a famous female ruler:

- Catherine the Great, empress of Russia in the eighteenth century, reputedly had a prodigious appetite for sex. Legend has it that she was killed by her bestiality practices. During one of her frequent conjugal visits with a horse, the rope sling that suspended the animal snapped, and the falling horse crushed the amorous woman. But the truth is that although Catherine had an appetite for sex, she did not indulge with her stallions. The rumor may have been started to undercut her claim to a place in history.

The stories you are about to read will one day fall into the domain of Historical Darwins, but don't wait! Enjoy them today while they're still fresh.

DARWIN AWARD: INTELLIGENCE BLUNDERS
Confirmed by Darwin
13 AUGUST 1999, MANILA

A deadly explosion in the Philippines' National Bureau of Investigation was initially considered to be a terrorist act. But the ensuing investigation linked the event not to criminals, but to careless NBI agents smoking near a bucketful of TNT. The blast killed seven people, including the perpetrator, and demolished the NBI Special Investigation Division. Grenades and other explosives also detonated in the fire. Officials are considering charging the division chief with criminal negligence for failing to safeguard seized explosives. But it is the perpetrator, crushing out his cigarette in a pail of explosives, who wins a Darwin Award.

Reference: Associated Press

DARWIN AWARD: HANGING AROUND JAIL
Confirmed by Darwin
2 APRIL 1998, WISCONSIN

Correctional institutions abound with "jailhouse lawyers" who will play any legal angle to improve their situations. Joseph, a twenty-year-old inmate of the Stevens Point Jail, planned a circuitous route to freedom. He would pretend to be crazy in order to be transferred to the minimum-security mental health facility, from which it would be easier to engineer an escape.

What would a crazy person do if he were trapped in jail? Joseph pondered the question, then decided to hang himself with a bed sheet until he was unconscious, while his bunk-mate alerted officials, who would cut him down and hopefully send him to the nuthouse.

Joseph's escape plan worked more quickly than he had anticipated. He hanged himself and was taken to the freedom of a grave the very next day.

Reference: *Louisville Eccentric Observer*

DARWIN AWARD: PEEPER PLUMMETS
Confirmed by Darwin
9 NOVEMBER 1999, MEXICO

A Mexican guard died from an excess of zeal while supervising an inmate's conjugal visit. Raul was closely watching his charge from the roof of the prison when he tripped over an air vent, crashed through the skylight, and fell twenty-three feet to land beside the bed where the inmate and his wife were, against all odds, enjoying an intimate moment.

The interrupted prisoner, offended by the intrusion, attempted to start a riot, which was squelched by prison security.

Prisoners in the Tapachula facility reported their jailor was in the habit of prowling the prison roof during conjugal visits, in search of prisoners to supervise. Local law enforcement officials reported that the guard was clutching a pornographic magazine, which was retained as evidence, and binoculars, whose sentimental value led to them being returned to the family of the deceased.

References: *La Crónica*, Reuters, *Le Journal de Montréal*, the *News* (Mexico City), InfoRed Radio

DARWIN AWARD: DEAD SPITTER
Confirmed by Darwin
15 JULY 1999, ALABAMA

A twenty-five-year-old soldier died of injuries sustained from a three-story fall, precipitated by his attempt to win a high-altitude spitting contest. He was so intent on victory, and so drunk, that he attempted to employ a dangerous and hitherto-untested technique. He backed away from the window, then hurled himself toward the metal guardrail while expectorating, in order to add momentum to his saliva.

In a tragic miscalculation his momentum carried him right over the railing, which he caught hold of for a fleeting moment before his grip slipped, sending him plummeting twenty-four feet to the concrete below. The military specialist had a blood alcohol content of 0.14 percent, impairing his judgment and paving the way for his opportunity to win a Darwin Award.

There is no report on the status of the payload he expelled into the night sky.

Reference: *Fort Hood Sentinel*

DARWIN AWARD: NEW DATING TECHNIQUE
Confirmed by Darwin
30 DECEMBER 1997, MEXICO

A security guard intending to impress female friends took a deadly gamble, losing his life in a game of Russian roulette at a La Paz fast-food restaurant. Police say Victor, twenty-one, died instantly on Saturday when he put his .38-caliber revolver to his head and pulled the trigger at a suburban hamburger outlet. Reports state that Alba was trying to "impress some female friends." We are certain that his bravado made a lasting impression on the women.

Reference: *Hoy de La Paz*, a daily newspaper in La Paz, Mexico.

DARWIN AWARD: RESISTANCE IS FUTILE
Unconfirmed by Darwin
1999

The U.S. Navy issues a safety publication that describes, as a prophylactic measure, injuries incurred while doing "don'ts." One story relates the fate of a sailor playing with a Multimeter in an unauthorized manner.

This sailor was a curious fellow. He wondered, what was the resistance level of the human body? Fortunately for him, he had the means to answer this question: a Simpson 260 Multimeter, familiar to most Navy personnel. It is a small unit powered by a nine-volt battery, which may not seem powerful enough to be dangerous—but that battery can be deadly in the wrong hands.

The sailor hooked up two probes and took one in each hand to measure his bodily resistance from thumb to thumb. The probes had sharp tips, and in his excitement he pressed his thumbs hard enough against the probes to break the skin. Once the salty conducting fluid known as blood was made available to the current, it traveled from fingertip to fingertip right across the sailor's heart, disrupting the electrical regulation of his heartbeat. He died before he could record his ohms of resistance.

The lesson? The Navy issues very few objects that are designed to be stuck into the human body.

References: U.S. Navy Safety Publication

DARWIN AWARD: WAIT FOR ME!
Unconfirmed by Darwin
1998

A trio of Marine officers went ashore to explore an overseas liberty port. In the wee hours of the morning, after a night of too much wining and not enough dining, they returned to their once-classy older hotel and began to drunkenly explore its highways and byways. They wandered aimlessly through the halls until one of them, who we'll call Curly, paused in front of two rococo wooden doors. He opened them and spied a delight: a dumbwaiter!

He alerted his companions, who we'll call Moe and Larry. "Check it out, a whatchacallit!" Sure enough, his buddies saw that it was indeed one of them whatchacallits, and a big one too. So big that a Marine couldn't help but wonder—if they scrunched up just a bit—two of them could squeeze inside and take it for a ride!

So they did.

Larry steadied the box while Curly crawled in and tucked his knees under his chin. Then Curly held the rope while Larry clambered aboard and settled into place. With a cheerful smile they released the rope, and with a whoosh they were gone!

Moe was astonished.

He lurched over to the opening, stuck his head into the shaft, and peered down through the darkness as the two men aboard the dumbwaiter fell to the earth like a ten-ton safe.

Kaboom!

They crashed into the bottom of the shaft, and their

little wooden box exploded in a shower of kindling and splinters that was quickly obscured by a dirty mushroom cloud of rising dust and detritus that had lain there undisturbed since the turn of the century.

In desperation Moe cried out, "Are you all ri—" But before he could finish his question, the counterweight, torn loose by the crash, fell from the top of the elevator shaft and cut his query short.

Reference: Navy Safety Center Summary of Mishaps

PERSONAL ACCOUNT:
INDUSTRIOUS BRAIN-DEAD PRIVATE
MARCH 1991

"Beam me up, Scotty!"

During the Desert Storm military operation there were relatively few casualties. Most of those that did occur were Darwinian in nature: put a few dummies around guns, tanks, grenades, and the like, and you can expect shit to hit the fan. Here is an example.

As with any modern conflict the use of land mines is critical. Desert Storm was no exception. Antipersonnel mines were dropped from airplanes by the truckful. These mines are small cylindrical "bomblets" that look fairly harmless. The notion of a harmless bomb was heightened in the tank unit, because tanks are virtually indestructible.

After one battle the 3AD 2/67th Armored Division occupied enemy positions that were brimming with bomblets, scattered like shells on the beach. Camp was set up in what seemed to be the middle of a minefield, proving that *military intelligence* is an oxymoron.

But the term *harmless bomb* is also an oxymoron. A tank accidentally ran over one, and it woke up everyone inside. Though there was no damage to the tank, the gun loader soiled his pants, and eyeball prints decorated more than one pair of glasses.

All the mines were marked with white cloth and chemical lights for safety, and soldiers were instructed not to tamper with them. The camp was a maze of white lines.

Then one industrious mechanic, Private Rock (PV1), decided that it was dangerous having all those mines lying around. So he dug a hole in the middle of the assembly area and proceeded to collect the mines and toss them into the hole. His activities went unnoticed until he had about twenty "harmless" mines, evidently duds, piled up in his hole.

Finally a sergeant hollered, "What do you think you're doing?!"

Private Rock responded, "I'm just getting rid of all these duds lying around." He threw one last mine in the pit, and the whole thing exploded. Private Rock was immediately beamed up by Scotty, leaving behind a twenty-foot crater and temporary hearing loss for half the company.

Luckily the shrapnel only caused flesh wounds to the surviving troops. For Private Rock's heroic efforts to eliminate pesky duds from the gene pool, he is posthumously nominated for a Darwin Award.

Reference: Personal account from a member
of 3AD 2/67th Armored Division.

Some military men argue that this story is woven from whole cloth. One lieutenant explained, "An enlisted member is only a PV1 for four months, long enough to complete basic and advanced training. PV1 Rock should have been promoted by the time he arrived in the Gulf. Besides, the word *rock* is military slang for *moron*, so the name PV1 Rock seems unlikely."

PERSONAL ACCOUNT:
5 SOLDIERS, 6 POLICE, 0 BRAINS
EARLY 1970S, NORTHERN IRELAND

An undercover military intelligence squad was patrolling a notorious Belfast area in plainclothes. After a long and perilous evening they emerged onto a York street and stopped for petrol and a few smokes. One of the soldiers asked the attendant if there was a pay phone, and the attendant pointed to the rear of the store.

As the soldier turned toward the phone, the attendant caught the flash of a concealed weapon. Alarmed and fearing a terrorist holdup, he vanished into the back room, where he phoned the local police station a hundred yards up the street. But instead of phoning the front desk, which knew about all military patrols in the area, he phoned a pal in the Criminal Investigation Department.

The CID was so excited by the thought of a good action going down, that they failed to consult with the local police at the front desk. They drove out, mob handed, to rescue their friend from terrorists.

The soldiers were just preparing to leave the petrol station when a car screamed to a halt across the street and disgorged six plainclothes policemen brandishing an assortment of weapons. Believing they were under attack by terrorists, the soldiers drew their own weapons, dived behind their vehicle, and opened fire. The police returned fire in earnest. For good measure an off-duty officer around the corner drew his weapon and fired four shots into the air.

The exchange lasted many minutes before a lone voice sounded, "Stop! Police."

Another voice shouted back, "Cease fire! Army."

Over one hundred rounds were fired across the busy intersection during the exchange. Not a single person was hurt, and the story was kept from the media to protect the identities of the "intelligence" officers involved.

Reference: Personal account from a lieutenant
who knew the participants well.

PERSONAL ACCOUNT: JET SKI JOCK
1999, CALIFORNIA

This week's Rocket Scientist award is taken from a U.S. Navy safety directive.

It was a quiet day at sea in the waters of southern California. On a ship forty miles from shore, a petty officer was on deck minding his own business, watching the waves go by, thinking lascivious thoughts. Suddenly, the sailor heard a mysterious disembodied voice calling, "Whiiich waaay to Catalinaaa?"

Startled, he looked around for the source of the eerie wail, expecting to find a frustrated ship's navigator. Instead he spied a jet ski racing alongside the ship. On the jet ski sat a civilian, steering with his knees, his hands cupped around his mouth, yelling desperately for directions to Catalina Island.

Jet skis are equipped with a key slot and a gas gauge as the full extent of their navigation equipment. Someone as dumb as this is a real rarity, even for the Navy, and it fascinated the sailor.

He decided to share his humorous good fortune with his pals, and called them over to the deck and pointed to the weird guy out in the middle of the ocean on a jet ski.

But the guy on the water scooter was intent on the sailor, and didn't notice him calling his friends. When the sailor gesticulated wildly on the flight deck, the man surmised that he was pointing out a vector to Catalina. So the jet ski took off for the island full throttle. The only problem is, the sailor wasn't pointing to Catalina at all. He was pointing to the dummy on the jet ski who was now headed for Pearl Harbor, Hawaii, hundreds of miles away.

It didn't take long for word of these goings-on to filter up to the bridge. As soon as the captain heard about the jet ski jock, he turned the ship around and gave chase. The crew ran down the wayward adventurer, dropped the tailgate, and encouraged him to putt-putt into the well deck.

Once aboard, the man dropped a bombshell. He had left a buddy out there somewhere, dead in the water. The last time he saw him, his friend was bobbing up and down in the ocean on a personal watercraft that was out of gas.

It sounds true, but did this event really happen? Random ocean searching is ineffective. The ship probably would have called in the Coast Guard to perform a faster and more efficient search with multiple vessels and helicopters.

It took all night, but the Navy ship finally rescued the running mate just after sunrise. When asked how he was feeling, he said, "I'm one happy dude." No, dude, you're one lucky puppy, that's what you are. You and your pal are proof positive that somebody needs to drop a few more chlorine tablets into the gene pool.

Reference: Anonymous personal account.

PERSONAL ACCOUNT:
NORTH PACIFIC DECKPECKER
NOVEMBER 1990, CANADA

Most sailor yarns are legends wearing the guise of truth. Some are deep-sea-blue embroidery, and others are downright lies. But this one is a true sea story, just the way it happened on a windswept gun deck not far from the open sea.

Like other vessels, the *H.M.C.S. Huron* was sent to the Arabian Gulf to endure two weeks of fires, floods, and famines. The ship returned to its home port of Esquimalt, British Columbia, where a subbie with an attitude joined the crew. Mr. Scarecrow, as he was called, was a typical know-it-all product of the Naval Officer Training Program. Within twenty-four hours he had succeeded in offending ninety percent of the lower deck, and the wardroom was not far behind.

One day Mr. Scarecrow paid a memorable visit to the foc'sle as two sailors were carrying out routine maintenance on the five-inch main armament called Tulio. Wandering to and fro unpleasantly, he complained about the state of the nonskid deck topping, which was scarred with rings exposing the metal below. Listening to his pointless tirade must have caused both seamen to momentarily throw discretion to the winds. His next comment was the final fuse. "Whatever could cause these unsightly rings?"

They explained to him that it was the work of the North Pacific Deckpecker. The Deckpecker is a large dark gray bird with nocturnal habits. It flies about the sea at night, searching for ships to land on. This particular bird feeds on the parasites that burrow into the ship's paint; the parasites in turn live on the cordite residues that accumulate about the gun decks. The rings on a warship's decks are caused by the Deckpecker pecking about its feet before moving to another position. Because of the bird's nocturnal habit and dark color, it is very rarely seen.

As the two salts described the Deckpecker, the officer became more fascinated at each revelation. Eventually, they even demonstrated the bird's call, a raucous sound that drew more crew to listen as the men contributed further details and corrected each other over minor points. When the impromptu lecture had come to a close, Mr. Scarecrow looked around at his rapt audience and said, "You know, I read about that somewhere."

Within the hour everyone from the greenest ordinary seaman to the captain knew the story. Of course, the real cause of the deck rings is the expended casings from the gun striking the deck. But the North Pacific Deckpecker lives on in sailor mythology.

Reference: William P. Sparling, Sr., personal account.

PERSONAL ACCOUNT:
FLACK VEST TESTING
1999

Many years ago, Police Explorers were shown a training video called *Flack Vest Testing by a Fool*. The starring role was held by a man who was goofing off and asking people to shoot his bulletproof vest.

While he was wearing it.

He took round after round of fire, from a .22 to a .357 magnum. The final shot was from an AR-15 rifle, cousin to the M-16. The bullet passed just a few millimeters below the Kevlar vest with its metal trauma plate. It entered the man's lower stomach, passed through his colon, and damaged his spine forever. He has trouble walking to this day.

Duh!

Reference: Anonymous personal account.

CHAPTER 7

Testosterone Poisoning: Macho Men

"If all else fails, immortality can always be assured by spectacular error."
—Observation by economist
John Kenneth Galbraith.

THE ISSUE OF OFFSPRING

In this chapter I have assembled a collection of stories whose common element is testosterone. The actions of the nominees are so shortsighted, so counter to common sense, that they can only be explained by the apparent need to strut superior risk-taking capacity. Such testosterone-inspired stories beg the question: What if these men have already reproduced?

Is a nominee automatically disqualified if he has offspring? Since genetic and environmental factors both play a role in determining our choices and behaviors, we will need to discuss each as a source of potential Darwin Award candidates, then attempt to answer the question posed above.

A concrete example will help illuminate the discussion. Imagine the sole reason a man wins a Darwin Award is because he has the hypothetical Explosive Stupidity gene, a gene that causes him to ignore the potential downside of

playing with bombs. The man who possesses this imagi-
nary gene tends to minimize potential dangers by rational-
izing that he is "good with explosives" and will not be
harmed. No matter how many hours of film footage he
sees showing flying body parts, and no matter how many
friends he knows who were injured in explosions, he will
never be convinced that he is anything but "good with explosives" and beyond harm's reach.

So one day he blows himself up playing Russian
roulette with a land mine, like the three fellows you'll read
about in "Fatal Footsie" (page 186), and his son is left to
bury the ashes.

The Explosive Stupidity son inherited half of his father's genes and half of his mother's. The son can be
thankful that he has only a fifty percent chance of possessing Dad's fatal Explosive Stupidity gene. Since children
have a good chance of *not* carrying a particular parental
gene, the presence of offspring will not disqualify the Explosive Stupidity man from winning a Darwin Award.

Genetic contributions, however, are only part of the
story. Our environment also plays a role in risk-taking behavior. This dichotomy is known as the "nature vs. nur-
ture" controversy, and professors regularly air competing
opinions on the subject. Let's see how environmental factors might figure into a Darwin Award.

If a child's father has the Explosive Stupidity gene, he
will learn from his father that it is okay to play with explosives. Even if the child lacks the Explosive Stupidity gene
himself, he will be more likely to win a Darwin because
he's conditioned to feel omnipotent around explosives. As
long as the father is around to encourage risky behavior,

the son's social environment makes it more likely that he will take the same dangerous risks.

But suppose Dad tosses a cigarette into a bucket of TNT like the detectives in MILITARY INTELLIGENCE: "Intelligence Blunders" and blows himself up. In that case it is highly unlikely that any child will follow in his footsteps. The environmental contribution is negated by the act that wins the Darwin Award. Again we are led to the conclusion that men who have reproduced are eligible to win a Darwin Award.

And finally, the child who inherits an unlucky gene will have his own shot at notoriety one day. So the rules do not disqualify nominees who have already reproduced.

The role of testosterone in reproduction is well documented, but it also plays a significant role in the manifestation of situational stupidity, as shown in the following eighteen tales.

DARWIN AWARD: JATO
1995 Darwin Award Winner
Debunked by Darwin

The Arizona Highway Patrol were mystified when they came upon a pile of smoldering wreckage embedded in the side of a cliff rising above the road at the apex of a curve. The metal debris resembled the site of an airplane crash, but it turned out to be the vaporized remains of an automobile. The make of the vehicle was unidentifiable at the scene.

The folks in the lab finally figured out what it was, and pieced together the events that led up to its demise.

It seems that a former Air Force sergeant had somehow got hold of a Jet-Assisted Take-Off unit. JATO units are solid-fuel rockets used to give heavy military transport airplanes an extra push for takeoff from short airfields.

Dried desert lakebeds are the location of choice for breaking the world ground vehicle speed record. The sergeant took the JATO unit into the Arizona desert and found a long, straight stretch of road. He attached the JATO unit to his car, jumped in, accelerated to a high speed, and fired off the rocket.

The facts, as best as could be determined, are as follows:

The operator was driving a 1967 Chevy Impala. He ignited the JATO unit approximately 3.9 miles from the crash site. This was established by the location of a prominently scorched and melted strip of asphalt. The vehicle quickly reached a speed of between 250 and 300 miles per hour and continued at that speed, under full power, for an additional twenty to twenty-five seconds. The soon-to-be pilot experienced G-forces usually reserved for dogfighting F-14 jocks under full afterburners.

The Chevy remained on the straight highway for approximately 2.6 miles (fifteen to twenty seconds) before the driver applied the brakes, completely melting them, blowing the tires, and leaving thick rubber marks on the road surface. The vehicle then became airborne for an additional 1.3 miles, impacted the cliff face at a height of 125 feet, and left a blackened crater three feet deep in the rock.

Most of the driver's remains were not recovered, however, small fragments of bone, teeth, and hair were extracted from the crater, and fingernail and bone shards were removed from a piece of debris believed to be a portion of the steering wheel.

Ironically, a still-legible bumper sticker was found: "How do you like my driving? Dial 1-800-EAT-SHIT."

This Darwin Award is the most popular of all time. Considered true for years, it was later confirmed as an Urban Legend by the Arizona Department of Public Safety. The story fooled the judges in 1995, so JATO has been grandfathered in as the 1995 Darwin Award Winner.

Officer Bob Stein of the Arizona Department of Public Safety talks about the JATO story. "I receive inquiries about accidents, drug busts, and investigations. About two years ago I picked up the phone and researched what has now become an Arizona myth. Even now I recieve about five calls a month from people wanting to know, did it really happen?"

Read the Official Arizona JATO denial.
www.DarwinAwards.com/book/jato.html

DARWIN AWARD: FATAL FOOTSIE
1999 Darwin Award Winner
Confirmed by Darwin
22 MARCH 1999, PHNOM PENH

Decades of armed strife have littered Cambodia with unexploded munitions and ordnance. Authorities regularly issue warnings to citizens, reminding them not to tamper with the devices.

Three friends recently spent an evening sharing drinks and exchanging insults at a local café in the southeastern province of Svay Rieng. Their companionable bickering continued for hours, until one man pulled out a twenty-five-year-old unexploded antitank mine that he had found in his backyard.

He tossed it under the table, and the three men began playing Russian roulette, each tossing down a drink and then stamping on the land mine. The other villagers, recognizing the inevitable, fled in terror. Minutes later their fears were confirmed when the explosive detonated with a tremendous boom, killing the three men in the bar.

"Their wives could not even find their flesh because the blast destroyed everything," the *Rasmei Kampuchea* newspaper reported.

References: *Rasmei Kampuchea*, Electronic Telegraph,
Reuters, (London) *Daily Telegraph*

Darwin Award: Guy Gulps Goldfish

1998 Darwin Award Winner
Unconfirmed by Darwin
29 January 1998, Ohio

Hungry or just stupid?

Wednesday was a fateful day for Michael. He was shooting the breeze with a group of buddies, watching a friend clean his fish tank, when the friend complained that one specimen in particular had become a fishy menace. It had outgrown the tank, and was eating other denizens of the aquatic community.

Michael volunteered to assist. He seized the five-inch fish and attempted to swallow it. Unfortunately, the fish continued its predatory ways by sticking in his craw. As he gasped futilely for breath, turned blue, and sank to his knees, his three friends realized that something was amiss. They phoned 911 and informed the dispatcher that Michael had eaten some fish, and was having trouble breathing.

Paramedics were quickly dispatched, and they arrived to find the fish tail still protruding from the victim's mouth. Despite their best efforts neither the fish nor the twenty-three-year-old could be resuscitated. The killer fish had claimed one last victim.

"If I dare you to jump off a bridge and you do it, you're stupid," Police Major Mike Matulavich said. Apparently Michael was not a victim, he was just another Darwin Awards contender.

DARWIN AWARD: DRY SPELL
Confirmed by Darwin
26 JULY 1991

Patrick lived to rue the day he planned a record-breaking 20-mile hike across the Badwater Salt Flats, the hottest place on earth. He completed 19.5 miles of his hike before collapsing on the scorching ground, never to rise again. Found with his body were a video camera and an empty three-quart water pouch.

The China Lake Rescue team located Patrick's parched body on his forty-first birthday, nearly two weeks after he set out on his desert hike. He was found only a half mile from his red Toyota truck, where gallons of fresh water waited on the seat. Patrick, a healthy 165-pound outdoorsman, had been dehydrated to 90 pounds by the blistering heat. What brought Patrick to such a sad state of desiccation?

Badwater routinely attracts extremists enticed by the lure of running a 150-mile course from Badwater to Mount Whitney, from the lowest point in North America to the highest point in the contiguous United States. Occasional brave souls attempt the one-way hike across Badwater to meet waiting friends and refill their water bottles. Only Patrick, our Darwin Award candidate, tried to make the trek alone with only three quarts of water.

According to District Ranger Mark Maciha, Badwater is consistently five to ten degrees hotter than nearby Furnace Creek, which registered a high of 134 degrees Fahrenheit in 1913. The summer sunshine heats the ground to almost 200 degrees, and the parched air approaches zero percent humidity. No rational explanation can be found for why this lifelong fitness fanatic failed to take sufficient water with him on his hike into this harsh climate. An estimated twelve quarts of water would have been required to survive the exertion of plodding through muddy salt.

Murder was ruled out by the autopsy, and suicide seems unlikely, as it was his third attempt to complete the trek. The most compelling theory is that he wanted to set the record for being the first man to make an unassisted round-trip hike across Badwater. A friend confides that he purposely kept rangers ignorant of his intentions because he knew they would watch over him.

And extra water is just so heavy!

Before his doomed hike, he boasted to several friends that he had calculated the exact amount of water he would need, and to save weight, he would take not a single drop more. In a lamentable miscalculation he carried only three quarts of water, which were simply insufficient to see him through to the other side.

Dr. Milton Jones theorized after the autopsy that Patrick may have sat down to rest with his truck within sight, but had lost so much body fluid that his heart was unable to pump the unnaturally viscous blood to his brain. He lapsed into unconsciousness and died.

Patrick was a healthy outdoorsman with an extensive knowledge of the desert. His father recalled, "He spent money on only two things: electronic equipment and going to the desert."

The video camera found by his body chronicles the first half of Patrick's hike before the batteries died. It ended with his haunting observation, "The only problem is that we have to hike back. . . . This is the real world. One false move, and you're dead."

Reference: *Los Angeles* magazine

DARWIN AWARD: SEQUINED PASTIE
Unconfirmed by Darwin
1998, NEW JERSEY

Burlesque clubs aren't as safe as they used to be. An unidentified twenty-nine-year-old man choked to death on a sequined pastie he had removed with his teeth from an exotic dancer at a Phillipsburg establishment. "I didn't think he was going to eat it," said a dancer identified only as Ginger, adding, "He was really drunk." If Ginger had used a stronger pastie adhesive, this Darwin winner would still be swimming in the gene pool.

DARWIN AWARD: GUN SAFETY TRAINING
Confirmed by Darwin
28 FEBRUARY 2000, TEXAS

A Houston man earned a succinct lesson in gun safety when he played Russian roulette with a .45-caliber semiautomatic pistol. Rashaad, nineteen, was visiting friends when he announced his intention to play the deadly game. He apparently did not realize that a semiautomatic pistol, unlike a revolver, automatically inserts a cartridge into the firing chamber when the gun is cocked. His chance of winning a round of Russian roulette was zero, as he quickly discovered.

Reference: *Houston Chronicle*

DARWIN AWARD:
THE WINNER GETS . . . A POSTMORTEM
Confirmed by Darwin
AUGUST 1999, AUSTRALIA

Drinking oneself to death need not be a lingering process.
Allan, a thirty-three-year-old computer technician, showed
his competitive spirit by dying of competitive spirits after
winning a Sydney hotel bar's drinking competition, known
as Feral Friday. The bar set a one-hundred-minute time limit
for alcohol consumption, and awarded points to drinkers on
a sliding scale. A beer was one point, and hard liquor was
eight points.

After bending his elbow for an hour and forty minutes of
hard drinking, Allan took the prize. He stood and cheered
his winning total of 236. "Winners never quit!" His high
score also netted him the literally staggering blood alcohol
level of 353 milligrams of alcohol per 100 milliliters of
blood, seven times greater than Australia's legal driving limit.

After several trips to the usual temple of overindulgence,
the bathroom, Allan was helped back to his workplace to
sleep it off, a condition that became permanent.

A forensic pharmacologist estimated that after downing
thirty-four two-point beers, four bourbons, and seventeen
shots of tequila, his blood alcohol level should have been
0.41 to 0.43 percent. But Allan had vomited several times after the competition ended, so his actual blood alcohol content was a bit lower at the time of death.

The cost paid by Allan was much higher than that of the hotel, which was fined the equivalent of $13,100 for not intervening.

It is not known whether Allan required any further embalming.

Reference: *Sydney Morning Herald*, Reuters

DARWIN AWARD:
I'M A MAN, I CAN HANDLE IT
Unconfirmed by Darwin
NOVEMBER 1997, PENNSYLVANIA

Ken, thirty-eight, was bitten by a cobra belonging to his friend after playfully reaching into the tank and picking up the snake. Ken subsequently refused to go to a hospital, saying, "I'm a man, I can handle it."

Falser words have seldom been spoken.

Instead of a hospital, Ken reported to a Jenkins Township bar. Cobra venom is a slow-acting central-nervous-system toxin. It works so slowly that he was able to consume three drinks while bragging that he had just been bitten by a cobra. He eventually succumbed to the poison, and died within a few hours.

DARWIN AWARD: WILLIAM TELL OVERTURE
Unconfirmed by Darwin
11 APRIL 2000, KENTUCKY

Larry and his friend Joseph decided to reenact the William Tell scene where the famous archer is forced to prove his prowess by shooting an apple off his son's head. But instead of apples, they used a beer can, which was closer to hand. Based on their rash choice of targets, you might suspect that the pair were teenagers, but in fact they were grown men of forty-seven. Larry put the beer can on his head and urged Joseph to shoot. And shoot he did, but Joseph was no William Tell. He missed the can, fatally wounding his lifelong friend Larry. Authorities said the men had been drinking, and that the shooting was not prompted by an earlier altercation in the parking lot.

Legends say the Swiss archer Wilhelm Tell was forced to use a crossbow to shoot an apple from his son's head by Bailiff Gessler, when he refused to pay homage to the symbol of the King of Habsburg. By the time the arrow found the apple, Wilhelm had already notched a second arrow aimed directly at the bailiff's heart, in case the first arrow should miss its mark.

DARWIN AWARD: SINK THE CUE BALL
Confirmed by Darwin
SCOTLAND

A twenty-three-year-old painter was known by his friends as "Death Wish" because of his reckless behavior, which included smashing glasses on his forehead and swallowing keys and glass. The police described him as possessing "good health physically but of low intellect." He was generally regarded as a bit of a fool, and probably resorted to such parlor tricks in an futile effort to increase his social standing.

Frequently, he would "swallow" a pool ball and then regurgitate it. The man had successfully performed this odd trick on many occasions, by keeping the pool ball at the back of his pharynx, or throat. This was possible because of the unique size of a pool ball.

One day, a typical day in many respects, he was seen consuming large quantities of draft lager. After closing time the publican readmitted him to continue illegal drinking with his friends. As the evening wore on, he was seen to place a cue ball in his mouth. He had done this so many times that his behavior did not cause any concern. But this time he found himself in difficulties. His friends tried to intervene, but he waved off their ministrations, ran out of the pub, collapsed in the street, and began to turn blue. Neither his friends nor an ambulance crew were able to save his life.

What went wrong?

On this occasion Death Wish had elected to swallow a cue ball instead of a pool ball. He wasn't aware that a cue

ball has a physical property that makes it perfectly suited for lodging in a pharynx. A cue ball is smaller than a pool ball so that it can be automatically recovered whenever it is potted. Unfortunately, our Darwin Award contender had never considered the mechanics of the pool table, and was unaware of this size disparity.

In the research article that describes this mishap, there is a photo of the victim's dissected throat, complete with the lodged cue ball and a rather ugly protruding tongue. For another riveting treatise on bizarre deaths, the author of the article recommends *Live Fishes Impacted in Food and Air Passages of Man* by E. W. Gudger, *Archives of Pathological Laboratory Medicine.* 2, 1962, pp. 355–375.

A cue ball is 4.75 centimeters in diameter, while pool balls are 5.03 centimeters. This small difference in diameter makes little difference in appearance. The problem for Death Wish was that a smaller diameter causes an exponentially lesser volume. A cue ball is 10.52 milliliters smaller than a pool ball.

In carrying out his farcical effort to prove he wasn't a fool, Death Wish disregarded common sense and simple mathematics, and lodged the cue ball in his pharynx.

Reference: "A Case Of Fatal Suffocation During An Attempt to Swallow A Pool Ball" by Gyan C. A. Fernando, MB, BS, MRCPath, DMJ, Forensic Medicine Unit, Department of Pathology, University of Edinburgh. From *Medicine, Science and Law*, the Official Journal of the British Academy of Forensic Scientists (1989) Vol. 29, No. 4, p. 308. Submitted by Grant Harris.

DARWIN AWARD: REPAIRS ON THE ROAD
Confirmed by Darwin
MARCH 1995, MICHIGAN

James was killed in March in Alamo as he was trying to repair what police described as a "farm-type truck." The thirty-four-year-old asked a friend to drive the truck on a highway while James hung underneath so that he could ascertain the source of a troubling noise. But James forgot to dress appropriately for the date. His loose clothing caught on a spinning bit of machinery, and his friend found James "wrapped in the drive shaft."

Reference: *Kalamazoo Gazette*

HONORABLE MENTION:
RIGHT TOOL FOR THE RIGHT JOB
Unconfirmed by Darwin
FLORIDA

Carl Wayne, twenty, was hit in the leg with pieces of the bullet he had fired at the exhaust pipe of his car. Apparently, while he was repairing the car he discovered a need to bore a hole in the tailpipe. When he couldn't find a drill, he used the tool of expediency and tried to shoot a hole in it with his gun.

Honorable Mention: Kiss Bites Back
Confirmed by Darwin
30 July 1999, California

Ken from Carlsbad accepted a dare and kissed a snake, landing himself in mortal danger. Ken's journey began when he proudly bragged to his friends that he had captured a deadly young rattlesnake the week before. They teased him by calling him a "snake lover," and they urged, "Kiss your girlfriend, Ken."

> Young rattlers can be more dangerous than older ones because they release the entire contents of their venom sac, and do not conserve any for subsequent strikes.

When he did, the three-foot rattler bit him on the lower lip and pumped its sac of venom into his face. His head and throat swelled to twice normal size, and emergency room personnel pumped vial after vial of antivenin into his bloodstream in a fight for his life. After three hours of intubation and twenty-five doses of antivenin, Ken was out of danger at the Tri-City Medical Center.

The swelling from snakebite can cause necrosis of the affected tissue, and Ken might have lost part of his face. He was fortunate, and will only see bruised and stretched facial skin in the mirror. But he will suffer the consequences of his foolish act for weeks, as flulike systems set in, caused by an immune response to antivenin.

Dr. Neil Joebchen, the emergency room physician, said, "In twenty-six years this is the worst case I've seen. His muscles were quivering like he had worms under his skin."

Reference: *San Diego Union-Tribune*

URBAN LEGEND: THE DOG AND THE JEEP
A Classic Urban Legend, one of the most popular of all time.

A fellow from Michigan buys himself a brand-new $30,000 Jeep Grand Cherokee for Christmas. He goes down to his favorite bar and celebrates his purchase by tossing down a few too many brews with his buddies. In one of those male-bonding rituals five of them decide to take his new vehicle for a test drive on a duck hunting expedition. They load up the Jeep with the dog, the guns, the decoys, and the beer, and head out to a nearby lake.

It's the dead of winter, and of course the lake is frozen, so they need to make a hole in the ice to create a natural landing area for the ducks and decoys. It is common practice in Michigan to drive your vehicle out onto the frozen lake, and it is also common (if slightly illegal) to make a hole in the ice using dynamite. Our fellows have nothing to worry about on that score, because one member of the party works for a construction team, and happened to bring some dynamite along. The stick has a short twenty-second fuse.

The group is all set up and ready for action. Their shotguns are loaded with duck pellets, and they have beer, warm clothes, and a hunting dog. Still chugging down a seemingly bottomless supply of six-packs, the group considers how to safely dynamite a hole through the ice. One of these rocket scientists points out that the dynamite should explode at a location far from where they are standing. Another notes the risk of slipping on the ice when running away from a burning fuse. So they eventually settle on a

plan to light the fuse and throw the dynamite out onto the ice as far as possible.

There is a bit of contention over who has the best throwing arm, and eventually the owner of the Jeep wins that honor. Once that question is settled, he walks about twenty feet out and holds the stick of dynamite at the ready while one of his companions lights the fuse with a Zippo. As soon as he hears the fuse sizzle, he hurls it across the ice at a great velocity and runs in the other direction.

Unfortunately, a member of another species has spotted his master's arm motions and comes to an instinctive decision. Remember a couple of paragraphs back when I mentioned the vehicle, the beer, the guns, and the dog? Yes, the dog: a trained black Labrador, born and bred for retrieving, especially things thrown by his owner. As soon as dynamite leaves hand, the dog sprints across the ice, hell-bent on wrapping his jaws around that enticing stick-shaped object.

Five frantic fellows immediately begin hollering at the dog, trying to get him to stop chasing the dynamite. Their cries fall on deaf ears. Before you know it, the retriever is headed back to his owner, proudly carrying the stick of dynamite with the burning twenty-second fuse. The group continues to yell and wave their arms while the happy dog trots toward them. In a desperate act its master grabs his shotgun and fires at his own dog.

The gun is loaded with duck shot, and confuses the dog more than it hurts him. Bewildered, he continues toward his master, who shoots at man's best friend again. Finally comprehending that this owner has become insane, the dog runs for cover with his tail between his legs. And the nearest cover is right under the brand-new Jeep Grand Cherokee.

Boom! The dog and the Jeep are blown to bits and sink to the bottom of the lake, leaving a large ice hole in their wake. The stranded men stand staring at the water with stupid looks on their faces, and the owner of the Jeep is left to explain the misadventure to his insurance company.

Needless to say, they determined that sinking a vehicle in a lake by the illegal use of explosives is not covered under their policy, and the owner is still making $400 monthly payments on his brand-new Jeep at the bottom of the lake.

PERSONAL ACCOUNT: ONE COOL DUDE
*Excerpt from a letter a physics student sent to his friends,
describing his senior year in college.*
JUNE 1998

My senior year of college opened with the customary research projects, grad school applications, and the like. But
that all changed two months ago. Some of you may have
heard rumors of some bizarre accident that I was involved
in. Here is the truth, unabridged, for those who actually
want to know.

In the second week of school the society of physics students held a roughly annual welcome-back party. As tradi-
tion dictates we made our own ice cream with liquid
nitrogen, 77 Kelvin, as a refrigerant and aerator. We spilled
a little liquid nitrogen onto a table and watched the tiny
drops dance around. Someone asked, "Why does it do that?"
That may have been the point of no return.

As is traditionally my role, I answered that the nitrogen
evaporates at the surface of the table, which creates a cushion of air for the drop to sit on, and thermally insulates the
drop, which minimizes further evaporation. That's why a
drop dances around without boiling, without touching the
table, and without spreading out like a pool of water.

Then I continued. I mentioned that the very same principle makes it possible to dip one's wet hand into molten lead,
or drink liquid nitrogen without injury. I had done the latter
several years earlier in a cryogenics lab, and remembered
the physics of how it worked.

Naturally those around me were skeptical. "It will freeze

your whole body. Remember *Terminator 2*?" But I was sure of myself. I had done it before, and I believed in the physics behind it. So I unhesitatingly poured myself a glass and took a shot. Simple. Swallow, blow smoke out my nose, impress everyone.

Within two seconds I collapsed to the floor, unable to breathe or indeed do anything except feel intense pain. The ambulance arrived. The police arrived. The journey to the hospital. The attempt to explain to baffled ER staff how something like this could happen. Then I passed out. I woke up the next morning connected to beeping machines. It turns out that, in accordance with popular belief, you really should not drink liquid nitrogen.

I subsequently learned a few things about liquid nitrogen. While you can safely hold it in your mouth and blow neat smoke patterns, you should never, ever swallow. The closed epiglottis prevents the gas from escaping, so expanding gas is forced into your body. And your esophagus naturally constricts around anything inside it, so even though there is a thin protective gas layer, your esophagus will manage to make contact with the liquid nitrogen.

I also learned that my memory was flawed. When I did the trick six years ago, I put it into my mouth and didn't swallow. Over time, the fine line between parlor trick and fatal accident must have blurred.

I was badly burned from epiglottis to stomach bottom. The gas expanded to fill my chest cavity, and the pressure collapsed a lung. During a grueling all-night surgery, they removed part of my stomach and ran my entire digestive

system on a machine. I was on a breather until my lung was restored. There are a few considerably uglier details which I will spare you. Doctors were impressed with my recuperative skills. I could breathe on my own after a few days. I could sit up in bed after a week, and was walking and eating in two. At eight weeks I'm virtually healed except for a number of unsightly scars.

And there's good news! I am the first documented medical case of a cryogenic ingestion. Read the *New England Journal of Medicine*. Three articles are in review now, and will be published soon.

My little adventure leaves me with a tendency to tell bad physics jokes at department meetings.

Reference: Anonymous personal account.

PERSONAL ACCOUNT: OUT OF THEIR HEADS
1997, HOLLAND

A group of employees were happy to escape work for a brief time and be bussed around on a tour by their company. It was a sunny day, and some of the more boisterous employees enjoyed sticking their heads out of a rooftop window. They were like puppies enjoying the wind in their ears.

The driver of the speeding bus told them several times to pay attention to the road and stop their foolishness. And then it happened.

Two men had their heads out of the window, singing as the wind blew across their faces, when the bus entered a viaduct. The cracking of bone was heard throughout the bus. Their heads did not come off cleanly, as you might expect, but the men fell dead into the bus with cracked heads and broken necks.

The chauffeur, asked whether safety regulations were properly observed, replied, "I always lock the damn thing when kids are in the bus, because kids just don't listen. But for God's sake, these were adults."

Reference: Anonymous personal account and
The Tonight Show with Jay Leno on NBC

PERSONAL ACCOUNT:
ROUND LAKE SHORTCUT
WINTER 1996–1997, MICHIGAN

To some Michigan residents drinking is a hobby, and drunken snowmobiling is considered family fun. The list of small-time Darwins grows, as drunks and idiots find novel ways of killing themselves on snowmobiles. Alcohol and high speed, combined with twisting tree-lined trails and thin ice, are guaranteed to bring out the inner moron.

Charlevoix is on Lake Michigan, and is divided north and south by a shipping channel that connects Lake Michigan with Lake Charlevoix. There is a smaller lake along the channel, called Round Lake, with a narrow opening into Lake Charlevoix that runs with fast currents year round. This narrow passage from Round Lake is notorious for having thin ice or no ice, even in the coldest depths of winter, because of the swift current.

Here's what happened to a pair of brothers.

Charlevoix has a festival in December that is the perfect opportunity to combine snow, alcohol, and fast sleds. The two brothers attended the festival on the north side of town, consumed a large amount of alcohol, and decided to go to another party on the south side. But instead of choosing from a multitude of longer, safer paths south, they elected to take a shortcut across Round Lake, during a driving snowstorm, through blowing slow, with very low visibility.

A short time later a woman walking her dog along the shoreline heard faint cries for help. Through a break in the snow she saw someone struggling in a hole in the ice.

She ran to the Coast Guard station and alerted the Ice Rescue Team, who were enjoying coffee and cookies following a successful training exercise.

They leapt to their feet and rescued the struggling man, who was hypothermic and virtually incoherent. He was rushed to the hospital, where the doctors began the process of warming and reviving him. As he regained consciousness, he looked around and asked, "How's my brother?"

Brother?

Back to the lake rushed the Ice Rescue Team. They found the two snowmobiles at the bottom of the lake, and a hole in the thin ice, but no brother. They searched for ten days, using divers and a video camera lowered through freshly drilled ice holes. Nothing. It was almost two years later when the brother finally turned up, still held together by his leather snowmobile suit.

As an example of stupidity with snowmobiles and alcohol, this one is a classic.

Reference: Anonymous personal account.

DARWIN AWARD: CHUTE BOY
Confirmed by Darwin
14 JULY 2000, CANADA

It was a dare that Sheldon, twenty-five, will literally never take again. He and a group of friends found themselves at a Calgary apartment after an evening in a local bar. It was there that a joking challenge was issued, possibly "Who wants to ride the in-house water slide?" The slide was actually a garbage chute and Sheldon volunteered. He tumbled into the opening, and his subsequent headlong slide beat the standard elevator service down to the first floor. An unforgiving trash compactor awaited his arrival, and there friends administered CPR until emergency crews reached the scene. But they were too late. The twelve-story fall had already dispatched Sheldon to his Darwinian demise.

Reference: Calgary Police Service Releases,
Ottawa Citizen, Calgary Herald

CHAPTER 8

Dangerous Liaisons:
Unsafe Sex

"Upon hearing about Charles Darwin's book The
Origin of Species, *the alarmed wife of the Bishop
of Worcester exclaimed, "Descended from the apes!
My dear, let us hope that it is not true, but if it is, let
us pray that it will not become generally known."*

EVOLUTIONARY HALL OF SHAME

The Darwin Awards in this chapter have been awarded
to people found in some surprisingly compromising
situations. Our species' obsession with sex is never more
apparent than when it leads to a particularly mortifying
public exposure. And our passion for sex leads us to other
activities that have an inimical impact on the gene pool.

To wit, the Evolutionary Hall of Shame Award given to
a California mother and son who were allegedly involved
in an incestuous relationship. Recently, the couple was discovered to have produced an infant daughter, and the
woman was pregnant with their second child.

Consanguinity has serious repercussions. When parents have a close genetic relationship, their children have
an increased incidence of genetic birth defects.

Authorities became aware of the peculiar relationship
when the woman's daughter told school officials she didn't

want to live at home anymore because her mother wanted her to call her brother Daddy. The forty-three-year-old woman and her twenty-three-year-old son had apparently been romantically involved for several years. The woman asserted to police that her live-in lover was not her son, but birth certificates confirmed the relationship.

Officials are pursuing the case because of the elevated potential for genetic disorders in children born to blood relatives. "They're having children and it looks like they're not going to quit anytime soon," said the deputy district attorney, who decided to pursue charges of felony incest. "That put me over the edge."

Many genetic disorders are recessive, which means that a healthy gene inherited from one parent masks the unhealthy gene inherited from the other parent. It is estimated that each one of us carries seven potentially lethal recessive mutations. We survive because our father and mother have dissimilar recessive mutations, and the "good" gene protects us from the "bad" gene.

A child born of mother and son, however, is much more likely to have twin copies of recessive genes, and suffer from serious genetic illnesses. That is why marriages between family members are considered taboo to most human cultures.

The perpetrators of the following escapades would have fared better if they had considered their peccadilloes to be taboo, as well.

Reference: San Francisco Chronicle,
San Jose Mercury News, Contra Costa Times

DARWIN AWARD: LOVE CRUSHED SEX
Confirmed by Darwin
JUNE 1999, FLORIDA

Okeechobee County investigators believe the death of Bryan, twenty-eight, was related to his wife's habit of stomping rabbits and mice for sexual pleasure. Stephanie, twenty-nine, was sentenced to two years of probation and community service for the death of her husband, who was found in a pit with a board over his body, crushed beneath the rear wheel of his sports utility vehicle.

Stephanie did not deny that she drove over her husband, but in her own defense she released tapes to the police showing her stomping small mammals to death. She was identified by a cryptic tattoo on her lower leg.

Such "crush flicks" are sold to people who derive sexual pleasure from the sight of death, especially at the hands of a woman. "It was abhorrent and cruel," said Assistant State Attorney Bernard Romero. "My first instinct was to seek the maximum penalty."

But Stephanie contended that she was an unwilling participant in the videos, and had been beaten many times by her husband prior to his bizarre death. Stephanie was charged in July with two counts of felony animal cruelty, which were later reduced to misdemeanors.

As for her husband, his death under the wheels of his car was presumably a loving sex act between consenting adults. But a man who would lie in a special pit while a woman he groomed for "crush flicks" drove over him, shouldn't be surprised when he winds up holding a Darwin Award.

Reference: *Fort Lauderdale Sun-Sentinel*, CNN

DARWIN AWARD:
BABY, YOU DRIVE ME CRAZY
Confirmed by Darwin
7 MAY 2000, ITALY

Full speed ahead! A young couple was killed in a freak car accident in Chieti. Germano and Franciska were discovered almost completely naked, and investigators presume they were having sex in their small Italian vehicle while it raced along Abruzzan roads at upwards of eighty miles per hour. Italian youngsters commonly use their cars for romantic trysts when parents forbid sex before marriage. But it is a mystery why this pair chose sex in a car traveling at high speeds over country roads. Germano lost control of the car in a bend, and the twenty-seven-year-old man and his twenty-year-old paramour were killed by the impact.

Reference: *Basler Zeitung, Corriere del Ticino*

Darwin Award: Sex and Suffocation
Confirmed by Darwin
21 March 1999, Bucharest

Romanian soccer midfielder Mario, twenty-four, and his friend Mirela couldn't wait to make love. As soon as their car was parked, they consummated their passion. They died from carbon monoxide poisoning shortly thereafter, inside the vehicle they had left running in the garage during their hasty liaison. The couple was discovered by Mario's father the following day. "They appeared to be unaware of the dangers of carbon monoxide," police colonel Dimitru Secrieru said.

9 May 1999, Mexico

A young Mexican couple was found dead in the back of a hearse. Jose, twenty-three, employed by the Pérez Díaz funeral home in Campeche, met Ana María for a romantic tryst in his hearse. He parked in a warehouse and left the engine running to provide air conditioning. In the enclosed location the carbon-monoxide-laden exhaust fumes seeped into the vehicle, fatally poisoning the couple. Their bodies were found when Ana María's mother initiated a search for her missing daughter.

Reference: Notimex, Reuters, Fox News

DARWIN AWARD: FATAL FLASHER
Confirmed by Darwin
16 DECEMBER 1997, TEXAS

A Dallas man who was exposing himself to passing traffic died during a performance one evening. Police were alerted to the roadside attraction by a motorist who had spotted Richard, forty-seven, standing naked on a railroad trestle. When officers arrived, the exhibitionist was taking a break under the trestle, still naked. As officers approached, he grabbed his clothes and ran back onto the railroad trestle. Richard leapt from the bridge, apparently aiming for a concrete support underneath, but missed and fell thirty-five feet to the ground. He died at Parkland Hospital an hour later.

Reference: *Dallas Morning News*

Honorable Mention: Chimney Manners
Confirmed by Darwin
9 May 2000, California

Shaun violated a restraining order when he climbed into his fifty-year-old paramour's chimney. But he could not violate the laws of physics. His body became wedged in the twelve-by-fifteen-inch shaft, and the thirty-year-old man was given a few hours of enforced solitude to contemplate his predicament.

Some time later a neighbor investigating a mysterious shrill voice followed its instructions and discovered the source to be Shaun, stuck Santa-like in the narrow shaft. "I couldn't believe anyone could possibly fit in there," she said. But there he was, twelve feet down the chimney in a squatting position, trapped with his arms above his head.

Los Angeles firefighters attempted to pull the man from the shaft with a rope, but were unsuccessful. Then they called in a jackhammer crew, who chipped a hole through the brick while the chimney dweller screamed in fear. Eventually the man was freed from his narrow confines, and found himself surrounded by reporters who had gathered during the lengthy rescue.

He triumphantly told the reporters, "I'm so stupid I'll probably win a Darwin Award!" Then Shaun was arrested on suspicion of stalking and burglary, and lost the liberty he had so recently gained. As he is still alive, his escapade doesn't meet our requirements for a Darwin, but Shaun will no doubt be proud to learn that he has received an Honorable Mention. Keep your eyes on this future winner!

Reference: *San Diego Union Tribune*

HONORABLE MENTION: BETRAYAL OF TRUSSED
Confirmed by Darwin
30 APRIL 1999, GERMANY

One night, firefighters rescued a wealthy German billionaire from a swiftly moving blaze that threatened his suite at the Hyatt Regency Hotel in the city of Cologne. "So what?" you might think. "That's a firefighter's job." But in this case they also saved the businessman from a kinky S & M situation straight out of the Marquis de Sade's recipe book.

The sizzling saga began shortly after midnight, when the tycoon hired a twenty-eight-year-old dominatrix named "Ramona" to bind, gag, whip, and otherwise humiliate him in his $400-per-night suite. Ramona was really no different from anyone else plying her trade. She was outfitted in fishnet stockings, chain-mesh bra, and six-inch spike heels. And she was packing the usual equipment: a cat-o'-nine-tails whip and other nasty sex toys. It was just another normal session of S & M sex, until a fire broke out in the adjoining room.

The conflagration began in the $1,700-per-night suite occupied by a computer mogul, who had been enjoying a bath in his sauna. Unaware of his neighbor's in-progress sex session, the man tore into the hallway and began warning the other hotel guests to flee.

Ramona heard the shouted warning about the fire, dropped her whip, and ran. She demonstrated an alarming lack of professional ethics when she left her client tied to the bed and unable to call for help. But her ethics were no less alarming than his lack of normal paranoia, in letting a stranger tie him to a bed.

When hotel workers and firefighters began a room-to-room search for endangered guests, they entered the German businessman's suite and stumbled across a bizarre scene that looked like something out of a porno flick. Firefighters stopped laughing long enough to fetch some cutting equipment to free the red-faced tycoon, while they continued to battle the flames.

The tycoon's clothing was evidently destroyed. He ran naked as a baby through the lobby and out of the hotel, with just a sheet to hide behind. "People cheered him on, which certainly added to his embarrassment," a fire department spokesman said.

Authorities identified the man as a "senior business executive" with a wife and two children, though perhaps for not much longer. They added that the kinky caper could have easily cost him his life.

Reference: *New York Post*

URBAN LEGEND: LIGHTNING DATE
1998, ARIZONA

A premed student from the University of Arizona was hoping to score with his date on a Friday night. To put the woman in the mood, he drove her to a secluded spot on Mount Lemmon, which overlooks the city of Tucson. They walked to an open knoll and admired the city lights.

Lulled by the romantic locale, the lissome lass succumbed to his passionate pleas. They tore their clothes off, made a bed of their garments, and began to make love. The heavy storm clouds rolling overhead mingled with the low rumble of thunder inside them. The excited lovers never looked up to see the charred skeletal remains of trees on the knoll.

Their idyllic clearing was a hotbed of electrical activity that night. With a blinding flash, a bolt of lightning struck the high point on the knoll, which happened to be the premed student's ass, and sought the path of least resistance straight down. Incredibly, he survived, albeit in excruciating pain.

The heat of the bolt had fused together flesh and latex so that the two lovers were now stuck together by their most intimate parts. The woman unfortunately did not survive the lightning strike. When the student looked down into the vacant eyes of his girlfriend and realized she was dead, his immediate repulsion caused him to jerk away from her. When he found that he couldn't, a wave of pain and nausea made him vomit into the girl's face and open mouth. The horror and pain of the situation caused him to black out.

Attracted by the smell, a bear made its way to the lovers and began to lick semidigested pizza and Buffalo wings from the dead girl's face. The student roused from his stupor. When he saw the bear, he realized that there was nothing he could do but remain silent, petrified with fear.

To his horror the bear became dissatisfied with just a lick and started to eat the girl, loudly crunching her facial bones just inches from his ear. The bear also sampled the student, scraping the back of his skull with its teeth, before moving on.

At 11:35 A.M. a group of hiking Girl Scouts arrived at the lovers' tryst, where the premed student's car was parked. Minutes later three shrieking girls discovered the student, who had regained consciousness several times in the night and had managed to drag himself and the partially eaten girl several meters toward the road. Doctors managed to separate the student from the corpse.

A hospital source reported that his penis resembled "a small piece of cauliflower" in its flaccid state. The first hint of arousal resulted in so much pain that the student was unable and unwilling to achieve an erection. Since his traumatized organ will no longer function in a procreatory sense, he is eligible for a Darwin Award.

URBAN LEGEND: GERBIL ROCKET

"In retrospect lighting the match was my big mistake. But I was only trying to retrieve the gerbil," Dick Grayson told the bemused doctors in the emergency room. Grayson and his partner, Tony Maloney, had been admitted for emergency treatment after a felching session had gone seriously wrong.

"I pushed a cardboard tube up his rectum and slipped Raggot, our gerbil, in," he explained. "As usual Tony shouted out, 'Armageddon,' my cue that he'd had enough. I tried to retrieve Raggot but he wouldn't come out again, so I peered into the tube and struck a match, thinking the light might attract him."

At a hushed press conference a hospital spokesman described what happened next. "The match ignited a pocket of intestinal gas and flame shot out the tube, igniting Mr. Grayson's hair and severely burning his face. It also set fire to the gerbil's fur and whiskers, which in turn ignited a larger pocket of gas farther up the intestine, propelling the rodent out like a cannonball." Grayson suffered second-degree burns and a broken nose from the impact of the gerbil, while Maloney suffered first- and second-degree burns to his anus and lower intestinal tract.

The gerbil's fate is uncertain.

Urban Legend: Hedonist Air Pumpers
16 April 1997

"The government must crack down on this disgusting craze of 'Pumping,' " a spokesman for the Nakhon Ratchasima hospital told reporters. "If this perversion catches on, it will destroy the cream of Thailand's manhood." He was speaking at a press conference held after the remains of thirteen-year-old Charnchai Puanmuangpak had been rushed into the hospital's emergency room.

"Most 'Pumpers' use a standard bicycle pump," he explained, "inserting the nozzle far up their rectum and giving themselves a rush of air, creating a momentary high. This act is a sin against God. But Charnchai took it further still. He escalated to using a two-cylinder foot pump, but even that wasn't exciting enough for him. He boasted to friends that he was going to try the compressed air hose at a nearby gasoline station.

"They dared him to do it, so under cover of darkness he sneaked in. Not realizing how powerful the machine was, he inserted the tube deep into his rectum, and placed a coin in the slot. As a result he died virtually instantly. Passersby are still in shock. One woman thought she was watching a twilight fireworks display, and started clapping."

"We still haven't located all of him," reported police authorities. "When that quantity of air interacted with the gas in his system, he nearly exploded. It was like an atom bomb went off."

"Pumping is the devil's pastime, and we must all say no to Satan," Ratchasima concluded. "Inflate your tires by all

means, but then hide your bicycle pump where it cannot tempt you."

The story is fake but the practice is real!

www.DarwinAwards.com/book/pump.html

Several clues highlight the implausibility of this story. It is impossible for methane in a person's rectum to explode when exposed to air. Ruptured colon from the air pressure, perhaps, but explosion, no. And a reader with a theological bent pointed out that a hospital spokesman in Thailand is unlikely to make references to God and Satan, which are not relevant to the largely nontheistic Buddhist population.

Urban Legend: Romeo and Juliet?
April 1999

Two students were in love and engaged. Unfortunately all of the parents involved disapproved of the marriage. The parents threatened dire measures if the students eloped. Caught in an impossible position of choosing between their love and their families, the students decided that they, like Shakespeare's Romeo and Juliet, would leave the world together.

Our Juliet told her friend, a pharmacist, that she was having trouble sleeping before exams, and asked for some potent sleeping pills. The pharmacist secured for her a small bottle of pills, plastered with warnings. "Danger! Use strictly as directed! Do not operate a moving vehicle!"

The two lovers locked themselves in a friend's dormitory room and tossed the key out the window. They shared a bottle of wine, made love, and then took the sleeping pills and kissed each other good-bye. Half an hour later they began to feel curious rumblings in their intestines. Soon they realized that Juliet's friend had given them laxatives, not sleeping potion!

There they were, locked in a small dorm room with the key ten floors below, and no toilets in sight!

The stench crept under the door and spread quickly throughout the building, alerting other residents. A security guard was summoned, who forced the lock and poked his face round the door. He quickly swung it shut, nearly overcome by the fumes. The unfortunate couple had to be rescued

by a SWAT team protected by gas masks. They were taken to the hospital and treated for severe dehydration.

It turned out that the friend at the pharmacy had been alarmed by the request for sleeping pills with no prescription. She contacted the parents, who conferred with one another and realized that something had to be done. Thus, the outcome: the marriage was belayed, both students were suspended from college, and both sets of parents were as "relieved" as their children.

A reader says the book *The Poisoning* by Russian writer Michail Weller closely resembles this story, except for his ending, in which the boy signs up for the army and leaves in shame while the girl slowly becomes close to the doctor who gave her the pills.

If you read Russian, you can read "The Poisoning" here:
www.DarwinAwards.com/book/russian.html

CHAPTER 9

Davey Jones' Locker: Watery Demise

"The problem with the gene pool is, there's no lifeguard."
—Stephen Wright aphorism.

DIVERGENT EVOLUTION

Do you think that idiots seem to be breeding more rapidly, and at an earlier age, than average? Do you think that geniuses seem to be breeding less frequently, and later in life, than average? What would happen to our population if that were the case?

Let's call the two sides of the equation the rabbits (notoriously fast breeders) and the pandas (slow breeders who have trouble reproducing in captivity). Imagine the animals are on opposite sides of a rope playing tug-of-war.

If a rabbit pair has five offspring by its twenty-fifth year, and a panda pair has three offspring by its forty-fifth year, what happens in a few generations? If you begin, like Noah's Ark, with two of each kind, in 225 years the pandas will have grown for five generations to a population of 15. The rabbits will have grown for nine generations to a hefty 7,600 members.

The descendants of the two panda bears will have trouble pulling the rope against so many rabbit opponents. How can the difference be so astounding? Is there something we missed?

You'll be relieved to hear that there *is* a missing factor. Remember we said the rabbits represent a stupid but rapidly reproducing population. If they're so stupid, they must be dying faster. Perhaps their death toll is so great that one in five rabbits leaps into danger and dies before reproducing. If that were the case, there would *still* be over 1,000 rabbits in 225 years. Clearly there is a strong tendency for nature to favor prolific breeders.

Years	Panda Children	Years	Rabbit Children
0	2	0	2
45	3	25	5
90	5	50	13
135	7	75	31
180	10	100	78
225	15	125	195
		150	488
		175	1221
		200	3052
		225	7629

Years	Rabbit Survivors
0	2
25	4
50	8
75	16
100	32
125	64
150	128
175	256
200	512
225	1024

If this model applies to humans, then the stupid and fast-breeding contingent must surely have overrun the species many generations ago. Perhaps that is why there are so many Darwin Award winners sailing toward their watery graves today, as in the following stories from Davey Jones' Locker.

DARWIN AWARD: HURRICANE HANGOVER
Confirmed by Darwin
15 AUGUST, 1969, ALABAMA

In 1969, Hurricane Camille claimed 143 victims along the Mississippi Gulf Coast. Most were guilty only of being in the wrong place at the wrong time, unlike twenty who perished while attending a beachfront "hurricane party" beer bash and barbecue.

Despite evacuation warnings delivered by vehement emergency teams, their festivities continued unabated. The partygoers defiantly declared that the concrete foundation and the second-floor location of their party provided plenty of protection from the impending hurricane.

Their confidence proved to be tragically misplaced when a twenty-four-foot wave slammed into the apartment, destroying the building and subjecting the partiers to gale-force winds and violent ocean surges. Most of these hurricane worshipers were killed. A few survivors were swept miles away, cheated of a Darwin Award by the capricious hand of fate.

Reference: Mobile (Alabama) *Press Register*

DARWIN AWARD: GONE FISHIN'
Confirmed by Darwin
25 MAY 1999, UKRAINE

A fisherman in Kiev electrocuted himself while fishing in the River Tereblya. The forty-three-year-old man connected cables to the main power supply of his home, and trailed the end into the river, producing an electric shock that killed the fish, which floated belly-up to the top of the water. The man had clearly demonstrated his understanding of the deadly effect of electricity, yet at the sight of all that tasty fish, he waded in to collect his catch without removing the live wire. The predictable result: He suffered the same fate as the fish.

References: Deutsche Press-Agentur, Bloomberg News Source

DARWIN AWARD: POLAR BEAR SWIM
Confirmed by Darwin
1 JANUARY 2000, CANADA

Believe it or not, there are people who dive into the ocean for a refreshing swim every New Year's Day. It's called a polar-bear swim, and it's just a crazy ritual to most of us. Anyone who has seen the film *Titanic*, or read a book about Eskimos, knows that icy water brings on rapid hypothermia and death. But our hero Adrian, studying for his doctorate in forestry, was not one to heed such trivial concerns.

This thirty-eight-year-old man was enjoying a hockey game with friends on Kingsmere Lake when he attempted a polar-bear swim between holes cut two meters apart on the lake. He dived in at 1:30 A.M., and failed to resurface.

It is common knowledge that it is nearly impossible to find a small hole in the ice once you've slid beneath the surface. Particularly when you are suffering from the effects of hypothermia.

Frantic friends jumped in but were unable to find him. They aimed car headlights at the hole to help Adrian

> A Scandinavian reader says, "The practice of swimming in ice holes is common in Finland, and tragedies occasionally happen without anyone questioning the general joy and positive health effects. It really is fun, and it's an excellent way to improve blood circulation and strengthen the heart. However, it is strongly advised never to *dive* into a hole, because you can easily lose the way out. And it is highly recommended that you avoid putting your head underwater. The scalp has the most temperature-sensitive skin, and it hurts!"

find his way back, but to no avail. "The water was only waist deep," said the man's brother. "He must have gotten disoriented."

Adrian's frigid body was recovered by firefighters, not far from the ice hole that tempted him to his doom.

Reference: *Toronto Sun, Ottawa Citizen, Montreal Gazette*

DARWIN AWARD: CRAPPY DRIVING AWARD
Confirmed by Darwin
9 OCTOBER 1999, MAINE

Some men die peacefully in bed, while others suffer less pleasant ends. Benjamin, twenty-three, lost his life in one of the most unappetizing manners possible when he careened into a four-hundred-thousand-gallon tank of raw sewage. Police speculate that he was driving his 1998 Mazda pickup much too fast to make the sharp right turn in front of the wastewater treatment plant.

He was apparently exceeding the speed limit by a generous margin, as his momentum carried him through a chain link fence, across an easement, and beyond a low post-and-rail fence surrounding the tank of decomposing sewage. Divers located his body beside his upright pickup on the bottom of the sixteen-foot-deep tank. The autopsy failed to provide a conclusive cause for death, but we speculate they will find he died from "taking too much crap."

3 MARCH 2000, PENNSYLVANIA

In a related event Andrew died in a messy farming accident at Crooked Creek Farm when he slipped into a manure spreader. Rescue crews failed to revive him (and who can blame them?). The cause of death was determined to be blunt-force trauma.

Reference: *Bangor Daily News*; WTAJ TV, Altoona, PA.

DARWIN AWARD:
CAN DUCK SHOOTERS SWIM?
Unconfirmed by Darwin
18 MARCH 2000, AUSTRALIA

The start of the Victorian duck-shooting season frequently ushers in a speedy reduction in the number of Australian duck shooters—and without assistance from the militant anti-duck-shooting lobby.

At the Cairn Curran Reservoir in central Victoria, a group of duck shooters set forth on a duck-shooting adventure in a small aluminum dinghy. This three-meter craft is termed a "tinny" for its cheap aluminum design. This particular tinny was rated to carry three adults.

Instead it was carrying Ringo and John and three friends, all from Melbourne. And it was carrying Ringo's son, six shotguns, and three crates of ammunition at twenty-five kilograms each. The tinny found itself loaded with over five hundred kilograms.

With all that gear and flesh onboard, there was no room for life jackets, so they were left behind in the car. Two men were wearing their waders, but waders act like lead weights if they fill with water, and it is virtually impossible to swim with them.

Always wear your life jacket. If this cautionary tale teaches you nothing else, let it teach you this.

Three hundred meters from shore the boat capsized, pitching its contents into the water. Three men were rescued

by boaters to live until another day's stupidity. John and Ringo were less lucky. They were found dead in their waders.

Sadly, the son, who was not at fault, also died.

Reference: Tom Croft, personal account, and the Melbourne *Herald-Sun*.

DARWIN AWARD: YOSEMITE HIKE
Confirmed by Darwin
10 JULY 1999, CALIFORNIA

A Yosemite hiker with sore feet stopped to cool his heels in the Merced River, where he slipped on algae-covered rocks and was swept over a 594-foot waterfall to his death. Siddiq was climbing Half Dome with three friends on Saturday when his lamentable choice of rest stops cost him his life.

Signs posted at the falls clearly state that if you go in the water, you will die. Not only are these warnings displayed in several languages, but they even show a stick figure falling over the edge.

But Siddiq paid no heed to the warnings. As he was carried over the Nevada Falls, his friends were already calling authorities for help from their cell phones. But help could not arrive in time. Rangers recovered Siddiq's body by helicopter a few hours later.

Siddiq is the fourth person to die at Nevada Falls in the last five years, park spokeswoman Christine Cowles said.

Reference: Associated Press

DARWIN AWARD: WET WILL HE
Confirmed by Darwin
23 AUGUST 1999, WASHINGTON

Rodney was jet skiing around Lake Washington, enjoying the sun and the power between his knees. After a few runs on the lake, he noticed that his battery was beginning to fail. He idled over to a dock near Juanita Beach Park and tied up his craft and ran to the car for his battery charger. When he returned, he plugged the charger into a 110-volt outlet and jumped onto his watercraft holding the booster cable.

Sizzle. He was found floating facedown beneath the dock later that evening.

Reference: *Seattle Times*

Darwin Award: Hard Work Rewards
Confirmed by Darwin
8 February 1999, Georgia

Fred of Forest Park was a forty-six-year-old plumber seeking employment. When he noticed a nearby sewer blockage, he didn't just sit around twiddling his thumbs. He leapt into action and used shovels to pry up a manhole cover, then entered the aromatic aperture. He was apparently trying to improve his chance of finding a job by demonstrating his plumbing skills. But in Fred's haste to identify the source of the sewer blockage, he neglected to set orange warning cones on the street around the open manhole. Upon exiting the sewer, he was struck by the undercarriage of an oncoming car, and was killed.

References: Associated Press; WSB TV, Atlanta, Georgia

HONORABLE MENTION: LOCH NESS MONSTER
Unconfirmed by Darwin
JUNE 1999, CALIFORNIA

Last summer at Lake Isabella, in the high desert east of Bakersfield, a woman was having trouble with her boat. No matter how she tried, she just could not get her new twenty-two-foot Bayliner to perform. It was sluggish in every maneuver, regardless of the power applied. She tried for an hour to make her boat go, but finally gave up and putt-putted over to a nearby marina for help.

A topside check revealed that everything was in perfect working order. The engine ran fine, the outboard motor pivoted up and down, and the prop was the correct size and pitch.

One of the marina employees jumped in the water to check beneath the boat. He came up almost choking on water, he was laughing so hard. Under the boat, still strapped securely in place, was the trailer.

References: Big Pig mailing list, *Deseret News Online*, Associated Press

URBAN LEGEND: DARWIN BEACH DEATH
11 OCTOBER 1998, AUSTRALIA

Woman drowned while performing fellatio.
A sexual romp at a popular Darwin beach ended in the death of a twenty-five-year-old woman, who drowned while performing oral sex on a man, the Northern Territory's Supreme Court heard. The woman had sexual intercourse with Mr. Payne, thirty-four, in "a number of positions" in the water off Pee Wee Camp beach, before she voluntarily submerged to perform fellatio on him.

Prosecutor Michael Carey told the court that while the woman was performing oral sex, "Mr. Payne became excited and put his hands on her head and kept her down there." The prosecutor said Payne told police that he noticed something was amiss when the woman stopped performing fellatio. He wondered what was going on, so he let her up.

"He says that she did not try to get up, she wasn't kicking or splashing, and that he really didn't do anything except let her up as soon as she stopped sucking on his penis," Prosecutor Carey told the court. He said that when Payne realized the woman was dead, he "freaked out," dressed, and drove away.

Payne, who has been in prison since two days after the drowning, pleaded guilty to committing a dangerous act. His counsel, Ms. Cox, told Justice Sir William Kearney that her client still had "recurring nightmares" about the drowning. "He keeps seeing it while he tries to sleep at night," Ms. Cox reported. She said a psychiatrist found that Payne had a deep

sense of shame about the incident. He required treatment for nervous outbreaks of boils twelve times in the past year.

Ms. Cox said that before Payne and the woman went into the sea, they had drunk eleven 750-millileter bottles of beer, and an autopsy found that the woman had a blood alcohol reading of .287—almost six times the legal Australian driving limit. "She might have just passed out under the water. That might explain why she didn't struggle," Ms. Cox told the court.

She said that although Payne had an alcohol problem, he was considered a quiet, shy, good-natured, and considerate person by his employers and friends. Ms. Cox said the unusual nature of the case meant there was no need for Justice Kearney to consider imposing a harsh penalty on him to deter others.

Justice Kearney sentenced Payne to four and a half years. "It's an unusual case that needed careful deliberation," Justice Kearney said.

An Australian reader commented, "Aussie girls stay down until the job is finished." Nevertheless, this story is classified as Urban Legend because of its improbable nature. A person held underwater drowns neither quietly nor quickly. Furthermore, Darwin area beaches are considered unswimmable by the natives. The names "Pee Wee Camp" and "Ms. Cox" are improbably scatological, and the name "Payne" (or pain) is too suggestive to be likely.

Reference: AAP Darwin Australia, *World Weekly News*

PERSONAL ACCOUNT: QUARRY STORY
JUNE 1985, VIRGINIA

Lou, a junior attending Christiansburg High School in Montgomery County, made headlines by drowning in a rock quarry. Nothing out of the ordinary, really, except for some unreported circumstances surrounding his death that were later revealed by his friends.

It turns out that as an end-of-school-year prank, Lou and three friends worked overtime one night, visiting dozens of their schoolmates' houses and stealing personalized license plates from their cars. One of the perpetrators estimates that they stole some fifty sets of plates.

The next day dozens of students called the school claiming they had missed the bus, and couldn't drive in because their license plates had been stolen. The school called the police, the police called an assembly, and at the assembly the police explained to the students that this was a felony criminal offense. A massive dragnet was going to be initiated.

Lou and his three friends quickly realized that they were in deeper water than they had bargained for. That night they took the license plates, hidden in a gym bag, to a nearby rock quarry and had a last bit of fun sailing the plates into the water one by one.

A couple of days later, after school was out, the boys decided to revisit the crime scene and knock off a case of beer. It was a warm, sunny day, and to their horror they saw that the entire quarry was agleam with the aluminum backsides

of dozens of license plates resting on the bottom. Lou quickly hatched a plan.

The quartet went back home and picked up a box of trash bags and an inflatable raft, then returned to the quarry. Lou grabbed a couple of forty-pound blocks, heaved them into the raft, filled a trash bag with air, and paddled the raft out to the part of the quarry where the plates were densest and most visible.

The plan was for Lou to grab the rock and the bag, roll over the side of the raft, quickly sink to the bottom, and turn the plates shiny-side down while breathing air from the bag. Then, presumably, he would drop the rock and use the bag to float himself back to the surface, and prepare for a second immersion if necessary.

In the end, however, Lou didn't need to go down but once. Because the water in the quarry was so clear, he had vastly underestimated the water's depth, which was about seventy feet. He rode the rock down until the water pressure burst the trash bag, but by then he was too deep to make it back to the surface.

The quarry story is the most controversial entry in the annals of the Darwin Awards. Would hand-carried rocks weigh enough to sink the bag of air? Would the bag really explode as it sank? Would a drowned body sink or float? Was the death *really* caused by thermal shock or decompression? And how could the sun's position in the sky have a bearing on the discovery of the license plates?

Read the saga on the Philosophy Forum.
www.DarwinAwards.com/book/quarry.html

By the time the dive team and other authorities arrived, the sun was setting and was no longer shining directly into the quarry. Lou's body was recovered and reported as a routine drowning death.

Nobody noticed the license plates.

Reference: Trey Howell, personal account,
and the *Roanoke Times* in Virginia

PERSONAL ACCOUNT: THE ICEMAN EXITETH
1 APRIL 1999, WISCONSIN

It is a common practice in Wisconsin, as in most places with cold winters and warm bars, to place an old car on the surface of a frozen lake and take bets on when the wreck will finally fall through the melting ice. With the spring thaw well underway, the betting season had mostly come and gone by the time Clinton, seventy-five, decided to test the ice for himself with an 861 International Tractor.

Clinton had been a farmer for fifty-eight years and had recently retired from the business. Perhaps he had intended to die as he had lived, on a tractor, or maybe the emissions from the vehicle had given him delusions of grandeur, the world will never know.

All was going Clinton's way that day. He had managed to drive his tractor onto the ice next to his boathouse, stand up in the bucket of his vehicle, and begin painting the boathouse. Then the ice, which happened to be covering thirty feet of water, gave way. Man and tractor plunged in, and Clinton didn't come out until the next morning, with the assistance of a dive team.

The headline read "Lake Claims First Victim," which is itself amusing considering the relatively passive role that the lake played in the drama. This is an example of natural selection based on a change of habitat. Charles Darwin would have said Clinton was a member of a species more suited to a field of corn than a thawing lake.

Reference: Anonymous personal account
and the *Ladysmith (Wisconsin) News*.

PERSONAL ACCOUNT:
RUB DUB DUB, MEN IN A TUB
DECEMBER 1997

A decade ago seventeen-year-old friends Ronnie and Steve were bored and scrounging about for something to do. They noticed that Steve's father had thrown out his old hot-tub to make room for the new one, and they decided to sail it across a nearby canal. The canal is a major shipping channel with a horrendous undertow. It is at least two hundred yards across.

Ronnie and Steve arrived at the canal, put the hot-tub in, and were pleased to see that it floated. They climbed in and managed to paddle a quarter of the way across the canal, but by this time quite a bit of water had splashed in. They decided that they needed to remove this water sloshing around their feet.

They reasoned, "Since the water is coming in over the sides, if we pull out this drain plug, the water will go out the hole." So they pulled out the plug, and you can imagine what happened next. Luckily it was a slow day in the shipping channel. The Coast Guard rescued them within an hour, freezing cold but unhurt.

Reference: Anonymous personal account.

PERSONAL ACCOUNT: SURPRISE FLUSH
1998, LONDON

An East London construction site was being converted from a school into private flats. The first phase had been completed, and one third of the school had been turned into flats, which were now occupied.

The builders were working on the middle third of the school, and needed to remove a large slab of concrete that formed the top landing of the central grand staircase. They decided that the best way to do so was to remove most of the slab supports, wait for all the flat occupants to go to work, clear the contractors, and remove the remaining supports, allowing the slab to drop directly to the basement.

The big day arrived. The main contractor checked to see that all of the cars had left for work, and he removed the manhole covers from the basement level. This was done to spare the site from the dust created by the downdraft of the plummeting slab, instead dissipating it into the sewers.

All nonessential contractors were instructed to stand clear. Laborers knocked on the flat doors to ensure that no residents were at home. Then the order was given, and like clockwork the slab crushed thunderously into the basement. The main contractor expressed his relief and glee at the smooth operation, the planning, and its total success.

A few moments later a gentleman in a soaking-wet, stained dressing gown approached and began to remonstrate with him. The man had stayed home that day with a case of diarrhea, and was perched on his toilet, which happened to be just on the other side of the wall from the basement.

At this point, recall that the manhole covers were open, and the slab, as expected, had acted like an enormous bicycle pump and blasted air into the sewers as it fell. Sewers which were, coincidentally, attached to this poor man's toilet drainpipe. The toilet water and its turgid contents were launched skyward in an arc described as a "font of cess" by the man in the dressing gown, who was drenched with the contents.

Reference: Anonymous personal account from a construction worker.

The "Valveless Flush Toilet," invented by Thomas Crapper, uses gravity and water pressure for its elegantly simple functionality. The system works on the sole condition that one does not flush while there is incoming pressure on the sewage line. According to a reader this principle became extremely well known in Seattle, Washington. A wooden sewage line was built that employed gravity to expedite waste to the ocean. The city is built on a hill by the coast of Puget Sound, perfect for Mr. Crapper's gravity-flow system, and it worked wonderfully at its unveiling—until they realized that the tide that whisked away the unwanted matter would also whisk it back, resulting in a prolonged stench. High tides covered the sewer outlet, creating an instant backpressure as raw sewage built up behind it. Citizens of Seattle made it a habit to post a copy of the tide table on the interior door of the lavatory, to help one decide whether to flush. If the tide table were not adhered to, a literal "fountain of cess" would ensue as the mounting city sewage pushed its way through one's plumbing.

PERSONAL ACCOUNT: TIDE-ALLY IMPAIRED
JULY 1999, ALASKA

The hero of this story lives on a farm in a small logging village in Alaska. It seems that this was a man who hadn't noticed, in all his years of living next to the port, that the tide comes in and the tide goes out. Not realizing

> A lifesuit is a vivid orange, insulated emergency suit used as a life preserver in very cold waters.

this, at low tide he tied his rowboat fast to a stationary object far out of reach, so that kids couldn't steal it for a joy ride. Predictably, the tide went out, the boat dangled for a time, the tide came in, and its binding sank the boat.

The man naturally wanted to retrieve his vessel, so he donned a bright orange lifesuit he had "borrowed" from a neighbor's skiff, and puttered out to his anchorage. Then he attempted to dive for his precious rowboat.

He didn't count on the buoyancy of the lifesuit.

Utterly frustrated, he brought the skiff back to shore, and placed two large rocks in the skiff. Then he returned to the site of his submerged treasure, tied the rocks to his waist, and jumped, apparently planning on cutting the ropes when he seized hold of his boat.

While sinking down, he realized that he had forgotten his knife on the boat, and was about to meet his watery demise. Luckily for him, and unluckily for us, by this time he had amassed an audience of five men, who were so astonished at his stupidity they weren't even sure what to do at first.

They eventually rescued him and he sold the boat.

Reference: Anonymous personal account.

CHAPTER 10

Man's Favorite Toy: Penis Envy

"Stupidity cannot be cured. Stupidity is the only universal capital crime; the sentence is death. There is no appeal, and execution is carried out automatically and without pity." A pronouncement by Lazarus Long in Robert Heinlein's Time Enough for Love.

LOSING THE FAMILY JEWELS

Here are thirteen stories of that rare species of Darwin Award winner, the survivor. Remember, death is not the only way to win a Darwin; losing the means to reproduce also qualifies one for a dubious victory. And it turns out that men are astonishingly prone to inactivating their favorite toy in a variety of chilling escapades. Once the genitals are gone, a man is unable to pass his reckless genes on to the next generation.

There are several objections raised to the notion of awarding a Darwin to an emasculated man. If he only loses his penis, his testicles can still produce sperm capable of fertilizing eggs in-vitro. If he loses his testicles but other cells remain, cloning techniques will soon progress beyond barnyard animals, and it will become difficult to permanently eliminate any person's genes, short of a jettison

into the sun. And who's to say that the man has not do-
nated to a sperm bank, where his genetic contribution is
frozen and dreaming of a willing incubator?

This chapter is about men who lose their ability
to procreate. But we would be wise to consider the
converse: A person who might be considered an anti–Darwin Award winner by adven-
titiously increasing his rate of reproduction.

Why are so few women winning Darwin Awards? Is it a conse-
quence of testosterone levels? Is it caused by one of the few
unique genes on the Y chromo-some? Or are women more dis-
creet in their fatal misadventures? Whatever the reason, male nominees predominate to an astound-ing degree.

A fertility doctor was recently convicted because he secretly do-
nated his sperm to his office's in-vitro fertiliza-tion efforts. What's the
harm in that? It is widely held to be unethical. The FIGO
Committee for the Study of Ethical Aspects of Human Reproduction states, "Members of the medical team involved
in the management of a recipient should not be donors."
But the doctor's excuse was that his sperm, being fresh,
were particularly effective at penetrating the eggs, and his
method resulted in a higher success rate than other, less-
endowed, fertility clinics.

Despite his lack of medical ethics, a reproductive entrepreneur such as this is far more effective than most of us
at ensuring that his opportunistic genes are generously
represented in the next generation. Therefore our doctor,
who blazed a broad reproductive swath, earns our first
anti–Darwin Award.

*   *   *

The recipients of the following Darwin Awards, on the other hand, have sorely curtailed their ability to contribute to the next generation.

HONORABLE MENTION:
ZANY NEW ZEALAND CONTEST
Confirmed by Darwin
7 JUNE 1999, NEW ZEALAND

A computer technician trainee set his own penis aflame in a successful attempt to win $NZ500 cash (about $250) and an equivalent bar tab.

Thomas stapled his penis to a white crucifix, poured cigarette lighter fluid over it, and set it on fire in his bid to win a controversial "How Far Will You Go?" promotion for Trader McKendry's Tavern in Christchurch. The event, sponsored by New Zealand Breweries, encourages patrons to compete for the most lewd act.

Thomas walked away with the top prize, which he used for car registration, a warrant of fitness, and registration for his bloodhound Puss. He obtained free medical treatment for his bruised and burned penis at a student clinic. He says his member "was a wee bit tender the next day," but after two weeks he is almost fully recovered, and expresses no regrets about his actions.

Photos of Thomas's qualifying demonstration:
www.DarwinAwards.com/book/penis.html

Thomas claims he is no masochist. "I'm a student so every bit helps. It was worth the money, and it's all better now. I thought my act was unbeatable." He intends to use the bar tab to buy burgers and pies every day for lunch.

Thomas's mother, who was in the audience, was pleased with her son's success. "He is a grown man and I'm relieved that he won. I would have hated for someone to go through all that and not achieve the object of it all."

References: *Sunshine Coast Daily* in Queensland, Australia; and the NZPA

HONORABLE MENTION: MR. HAPPY'S VACUUM
Confirmed by Darwin
13 MAY 1998, NEW JERSEY

There's apparently not much to occupy residents of Long Branch during the warm May evenings. A fifty-one-year-old man decided to satisfy his fantasy of robotic love by seeking sexual gratification with his vacuum cleaner. Most men would think twice before poking a valuable organ into a vacuum, but this optimistic fellow had no safety qualms, and besides, using a vacuum cleaner had the appealing aspect of tidying up his mess afterward.

Our horny hero didn't realize that the suction on his handheld Singer A-6 was created by a blade whirling just beneath the hose attachment, adjacent to the collection bag. His search for pleasure was cut short seconds after he stuck his penis into the vacuum and the blade lopped off part of his glans. With a sense of loss he staggered to the phone and called police. He told them he had been stabbed in his sleep.

Events such as this are common in Germany. A graduate dissertation at the University of Munich details many of such injuries, and includes case studies and interviews with the involuntary volunteers. Those interested can read *Penisverletzung bei Masturbation mit Staubsaugern* Theimuras, Michael Alschibajy, von der Universität München.

When police pointed out suspicious evidence, the victim claimed not to remember the incident.

Surgeons at Monmouth Medical Center stopped the

bleeding, but were unable to reattach the half-inch severed part. Though the man is still alive, his ability to reproduce has been curtailed by both his injury and his proclivity for household appliances.

References: Associated Press, *USA Today*, UPI, *Wausau Daily Herald*, KROG Los Angeles, *The Star-Ledger* (New Jersey)

DARWIN AWARD: PRIAPISM TAKES A PENIS
Confirmed by Darwin

Doctors warn of a dangerous new method of cocaine abuse: injecting the drug directly into the urinary tract. Physicians from the Cornell Medical Center reported the case of a thirty-four-year-old man who suffered severe bleeding under the skin after pumping cocaine into his urethra. It led to complications that destroyed his penis, nine fingers, and parts of his legs.

"They fill an eye dropper or a syringe with a cocaine solution and inject it into the penis," said Dr. Samuel Perry, a professor of clinical psychiatry.

The man had injected cocaine before intercourse in an effort to enhance sexual performance. He was admitted to the hospital because his penis had remained erect for three days, resulting in a painful inability to urinate. The medical term for a prolonged erection is *priapism*.

On his third day in the hospital the man's erection suddenly subsided. Over the next twelve hours blood leaked into the tissues of his feet, hands, genitals, back, and chest. The blood then coagulated, causing tissues to die over large areas of the patient's body. He was transferred to the burn unit of New York Hospital.

Doctors there were forced to amputate the man's legs and all but one finger, to stop the spread of gangrene. The patient's penis fell off by itself. The man is currently recovering in a rehabilitation facility.

Drug abuse treatment experts have previously reported external use of cocaine as a sexual stimulant. Cocaine powder is rubbed onto the surface of the genital organs by both men and women in an effort to halt premature ejaculation or improve sexual sensations. Men who inject cocaine into the penis claim that it gives them a sexual high.

"We report this case to alert clinicians to this new method of cocaine abuse and to describe its rare and previously unreported complications," the doctors concluded.

Reference: "Intraurethral Cocaine Administration"
J. C. Mahler, S. Perry, B. Sutton, *JAMA* 1988 Jun 3;259(21):3126

DARWIN AWARD: MAN SLICES OFF PENIS
1997 Darwin Award Winner
Confirmed by Darwin
5 DECEMBER 1997, CALIFORNIA

Alan, forty-eight, was found collapsed on the front lawn of his brother's Fairfield home eight hours after his penis had been cut off at the base. Paramedics rushed him to North Bay Medical Center, where surgeons were unsuccessful in their attempt to reattach his severed organ.

Alan initially blamed the maiming on a woman named Brenda, whom he'd met a local gas station the previous night. He claimed he had brought Brenda to his trailer, parked in the driveway of his brother's Fairfield home, and had sex with her. Around 3:00 A.M. he woke to find the woman ranting about revenge, and she cut off his penis with a razor-sharp hobby knife. She then fled the trailer on foot, leaving his penis behind.

Details of the attack were sketchy, and police were unsure why Alan could not defend himself.

A heated manhunt for Brenda ensued. Meanwhile, after being discharged from the hospital on Monday, Alan hitched a trailer to his pickup truck, drove off, and disappeared. Detectives were eager to interview him again, but were unable to locate him due to his transient lifestyle.

More intriguing details began to emerge.

Alan was arrested in the 1970s for drug possession and drunk driving. In 1982 he was jailed for taking his young daughter out of state. His ex-wife described him as a packrat who enjoyed taking trips in his mobile trailer home.

In 1983 Alan had been convicted of voluntary man-slaughter of a twenty-three-year-old Suisun City woman found strangled in a car parked at a local Denny's restaurant. Alan confessed to the murder, saying that she taunted him about his inability to achieve an erection when he tried to have sex with her. His statement was ruled inadmissible due to improper police interrogation techniques, and prosecutors agreed to let Alan plead guilty to voluntary manslaughter. He served half of a six-year prison term.

Police speculated that the woman who cut off his penis might have been carrying out a fourteen-year-old vendetta for the slaying of her friend. But the truth was even stranger.

When Alan was finally located and interviewed, he admitted that he cut off his own penis! A voice stress analyzer indicated that he was telling the truth. "At this point there is no evidence that a crime occurred," Police Lieutenant William Gresham said in a press release. "The case is being reclassified as an injured person report." Alan may face misdemeanor charges for filing a false police report.

Ironically, Alan works as a pipe fitter, according to court records. Now he has one less pipe to play with.

Reference: *San Francisco Chronicle, San Jose Mercury News,*
Contra Costa Times

DARWIN AWARD:
LOVE FROM THE HEART
Unconfirmed by Darwin
MARCH 1998, TENNESSEE

A teenage Knoxville boy read in an adult magazine that you
could hook a cow heart up to a battery and create an organic sex toy. Thinking to improve on the original model,
he hooked it up to the household current, electrocuting
himself and setting fire to his house.

1997, ITALY

A man was found naked and dead with an unidentifiable
mass attached to his penis. The coroner examined the man
and, in a brilliant display of detective work, determined that
he had connected a cow heart to electric cables and plugged
the apparatus into a normal 220-volt outlet. He then tried to
have sex with this quickly pumping toy, and was killed by
the electricity unleashed by the object of his desire.

Reference: *X-Factor* magazine, *Hobby & Work* magazine, *Fortean Times*

HONORABLE MENTION:
SCOUTMASTER SNARE
Confirmed by Darwin
1999, FLORIDA

A Boy Scout leader in search of a new adventure was found after being lost and lonely for thirty-six hours. Hank was reported missing by his wife after he failed to return from a bicycle trip. More than forty deputies searched on foot, by horseback, and with helicopters, until the Scout leader's cries for help led to his discovery—hanging naked upside-down from a tree. A video camera was positioned to capture Hank's adventures on tape, which undoubtedly did not turn out at all as he had planned.

Investigators speculate that the man, who wore only his shoes and was found suspended twelve feet off the ground by a rope tied around his ankles, was attempting to film an "autoerotic situation." After completing his solo ritual, Hank was too tired to pull himself back up and untie his feet, so he hung upside down until rescued. Lack of circulation led to such severe injuries to his feet that the left extremity had to be amputated.

An officer of the local Sheriff's Department observed sagely, "Some things are better confined to your own home."

Reference: Associated Press, InsideCentralFlorida.com, MSNBC.com, *Fort Lauderdale Sun-Sentinel*, APB News

HONORABLE MENTION:
SCROTUM SELF-REPAIR
Confirmed by Darwin
1991, PENNSYLVANIA

This is a true story based on a medical report, and contains disturbing descriptions. Warning! Read at your own risk.

One morning a doctor was summoned to the emergency room by the head nurse, who directed him to a patient who had refused to describe his problem more specifically than to request "a doctor who took care of men's troubles." The patient was pale, feverish, and obviously uncomfortable, and he had little to say as he gingerly opened his trousers to expose a bit of angry red and black-and-blue scrotal skin.

The nurse left the two men in private, and the patient permitted the doctor to remove his trousers and two or three yards of foul-smelling, stained gauze wrapped around his tender scrotum, which was swollen to the size of a grapefruit. A jagged laceration, oozing pus and blood, extended down the left scrotum.

Amid the matted hair, skin, and pus, the doctor spotted some half-buried dark linear objects, and interrogated the patient on their identity. Several days earlier, the man allowed that he had injured himself at work in the machine shop, and he had closed the laceration with a staple gun. The linear objects were one-inch staples used to mount wallboard.

The medical staff X-rayed the patient's scrotum to locate the staples, and gave him tetanus antitoxin, broad-spectrum antibiotics, and a hexachlorophene Sitz bath prior to surgery the next morning.

Eight rusty staples were excised. The left testicle had been torn off and was missing, but the stump of the spermatic cord was recovered and stitched shut. Convalescence was uneventful, and before his release a week later, the patient confided his story to the doctor.

An unmarried loner, he rarely lunched with his coworkers. Finding himself alone at noon, he had begun the regular practice of masturbating by holding his penis against the canvas drive-belt of a large piece of machinery. One day as he approached orgasm, he lost his concentration and leaned too close to the belt. His scrotum suddenly became caught between the pulley wheel and drive belt, and he was thrown into the air, landing a few feet away. Unaware that he had lost his left testis, and too traumatized to feel much pain, he stapled the wound closed and resumed work.

We can only assume he abandoned this method of self-gratification.

Reference: Medical Aspects of Human Sexuality, July 1991

HONORABLE MENTION:
HORSE DRUG EXPERIMENT
Unconfirmed by Darwin
1999, GUYANA

The condition of unusually long periods of penile erection is called "priapism." This unfortunate condition results when blood is unable to drain normally, as it does in a flaccid penis. Priapic erections can be caused by blood disorders, such as sickle-cell anemia or leukemia, but sometimes stem from an inexplicable application of stupidity on the part of the owner.

A Guyana newspaper carried the story of one bozo who, wishing to prolong an upcoming mattress rendezvous with his girlfriend, overdosed on Cantarden, a drug used for putting horses in heat. The drug had its intended effect, and the man discovered to his horror that his penis had become painfully erect and refused to go down.

After days of agony he reluctantly sought medical assistance, hoping he could keep the whole thing confidential. Luck was against him. When word of his erection slipped out, he became the town laughingstock, and adding insult to injury, he was nicknamed "Staff Sergeant" by the locals.

After three weeks of humiliation and misery he underwent surgery to relieve the condition. Now his only problem is that he can't get it up. Take lesson from this man's story, and leave drug experimentation to the professionals.

URBAN LEGEND:
FROG GIGGIN' ACCIDENT IN ARKANSAS
25 JULY 1996, ARKANSAS

Two local men were seriously injured when their pickup truck left the road and struck a tree near Cotton Patch on State Highway 38 early Friday morning. Woodruff County Deputy Dovey Snyder reported the accident shortly after midnight.

> "Frog giggin' equipment consists of two items: a gaff hook and a pair of waders. Night hunters creep into the creeks and swamps of the rural South, listening for the croak and the plop of a bullfrog. Grab a frog with the hook, fill up a pailful, and you've got yourself some French gourmet frog legs." Wisdom from an Arkansas native.

Thurston Poole, of Des Arc, and Billy Ray Wallis, of Little Rock, are listed in serious condition at Baptist Medical Center. The accident occurred as the two thirty-something men were returning to Des Arc after a frog-giggin' excursion.

On an overcast Sunday night Poole's pickup truck headlights malfunctioned. The two men concluded that the headlight fuse on the older model truck had burned out. A replacement fuse was not available, but Wallis noticed that the .22 caliber bullet from his pistol fit perfectly into the fuse box next to the steering wheel column. Upon insertion of the bullet the headlights again began to operate properly and the two men proceeded eastbound toward the White River Bridge.

After they had traveled approximately twenty miles, just before crossing the river, the bullet apparently overheated, discharged, and struck Poole in the right testicle. The vehicle swerved sharply to the right, exiting the pavement and striking a tree. Poole suffered only minor cuts and abrasions from the accident but will require surgery to repair the bullet wound. Wallis sustained a broken clavicle and was treated and released.

"Thank God we weren't on that bridge when Thurston [shot his intimate parts off] or we might have been dead," stated Wallis.

"I've been a trooper for ten years in this part of the world, but this is a first for me. I can't believe that those two would admit how the accident happened," said Snyder.

Upon being notified of the wreck, Poole's wife asked how many frogs the boys had caught.

This story was incorrectly attributed to the *Arkansas Democrat Gazette*, which denied ever publishing the story, and issued a decisive denunciation an October 17, 1996. "From all indications it's not true. There is no town of Cotton Patch in Woodruff County. No Deputy Snyder has ever worked for that county's Sheriff Department. And attempts to verify the existence of Wallis and Poole have been futile."

Read the full denunciation!
www.DarwinAwards.com/book/arkansas.html

PERSONAL ACCOUNT: JUMP ROPE BLUES
1999

While working in the operating room as a surgical technician, I had occasion to assist a surgery on a guy who was ex-
periencing extreme pain in his lower abdomen. When we
disrobed him to prep him for surgery, we noticed the tip of a
round blue object sticking out of his urethra. The doctor
palpated his bladder and determined that there was definitely something odd in there, so we opened up his pelvis
and found his bladder bulging with a tangled mass of blue.
When we opened his bladder, a serious procedure as they
tend to leak after being stitched or stapled shut, we extracted about six feet of knotted nylon jump rope.

The guy had evidently decided to cut the handles off the
rope and slide it into his urethra. Once he worked a foot or
two of the semirigid coil into his bladder, it naturally began
to unwind and twist into
a convoluted mass. The
end of the rope slipped
through a coiled loop,
and when he tried to pull
his exercise equipment
out of his bladder, the
coils tightened around the
free end and created a
huge knot. Try it yourself—most fishermen know several
knots that tie by coiling a line and passing the bitter end
through its own loops.

Lessons to Live By:

1. Using a jump rope is not always healthy.
2. Misuse of exercise equipment
   can result in serious injury.
3. Things should exit, not enter,
   via the urethra.

Needless to say, the guy paid dearly for his little experiment in autoerotica, and probably will never experience the joy of a jump rope in quite the same way.

Reference: Anonymous personal account.

PERSONAL ACCOUNT: DISCO DORK
JUNE 1999, UTAH

Paramedics were called to a discotheque in Salt Lake City where a young man had lost consciousness on the dance floor. Bystanders reported, "One minute he was dancing, and the next minute he was lying on the floor turning blue." His skin was pallid from lack of oxygen. The paramedics concluded that the man had suffered a heart attack, and loaded him into the ambulance. He died en route to the hospital.

In the emergency room the true cause of death was discovered when staff removed his personal effects. It turned out that he had strapped a roll of quarters to his crotch in the hopes of making his equipment appear larger. Unfortunately, the quarters were tied with surgical tubing, which had cut off circulation to his leg. Apparently the lack of blood flow, combined with the exertion of dancing, triggered his heart attack.

The moral of the story is: Size does matter. If his brain had been larger, he'd still be alive.

Reference: Copyright © 1999–2000 Gary K. Sloane,
personal account. Used by permission.

PERSONAL ACCOUNT: BRIDGE BOWLING
OCTOBER 1999

There are jerks, and then there are real jerks, like people who think it's funny to drop heavy objects on cars from a highway overpass. Hector, our hero and Darwin contestant, is one of the real jerks.

Late one night he found the perfect spot for his hobby. An old overpass had been torn down, and its replacement was almost finished. Hector located a boulder of concrete identifiable as a piece of the old bridge by the rusty steel rods protruding from it. Using the rods as handles, he dragged the mass onto the new bridge, and right to the brink of the fast lane below. Then he heaved it onto the edge of the low construction fence, a position so precarious that he was barely able to hold it in place. There he lurked, waiting for a victim.

Hector wasn't aware of the hook-shaped, rusty metal rod that was dangerously close to his private parts.

When the next car came, he timed the moment just right, and rolled the concrete over the brink. The sharp metal rod pierced his jeans and hooked itself firmly in his flesh, and before our hero could react, the boulder began to painfully pull him off the overpass.

Panicked, he managed to grab the handrail and hang on for dear life. Unfortunately, neither the jeans nor his delicate flesh could support the weight of the falling boulder. The metal rod ripped through flesh and fabric, and tore his jeans from his body, pulling them down until his feet got stuck in the cuffs. At this point his hands slipped from the rail, and he fell to the roadway, suffering more injuries and knocking himself out as he landed on the roadbed on top of the concrete boulder.

The driver of an oncoming car braked and swerved as the boulder and the man crashed into her headlight beams. She avoided the man lying on the road, but not the boulder, which became stuck in the bumper of her car. The man was still attached to the boulder by the jeans at his feet. He was dragged ninety feet, until the car finally braked to a halt.

After this experience our hero's contribution to the gene pool was in rather poor shape, as you might imagine, partially hanging in shreds from the boulder, partially spread across thirty yards of highway. Heroic doctors managed to save quite a bit of him, almost spoiling his chances for a Darwin. In the nick of time an infection caused by the rusty wound finished the job, resulting in the loss of his private parts.

As a bonus to society, as soon as he was able to leave the hospital, he was locked away from the public for a long time. Not only did he lose his ability to contribute to the gene pool, but he also will no longer be able to disturb the survival of the fittest by throwing heavy objects at them.

Reference: Anonymous personal account.

PERSONAL ACCOUNT:
PISSING INTO THE WIND
1998, ARIZONA

A man from Tucson was visiting Windy Point, overlooking a sheer cliff on Mount Lemmon. Because so many gawking tourists flock to the site, a wrought iron safety fence had been installed to prevent hapless rubberneckers from slipping off the cliff.

The weather was terrible.

Thunderstorms are common in late summer, and the Tucson man was the only one there. He decided to take advantage of the privacy and urinate through the bars of the fence over the cliff face below.

As he urinated, lightning from a powerful desert thunderstorm struck the fence, which was a perfect lightning rod due to its size, location, and composition. The charge traveled through the fence along the path of least resistance, not only to the ground but also up the man's urine stream, causing his penis to explode.

Reference: Anonymous personal account.

CHAPTER 11

Foolish Ingenuity: End of the Line

"Think of it as evolution in action."
—Quote from a sign above a
popular suicide jump spot in the
Larry Niven and Jerry Pournelle
book, *Oath of Fealty.*

EVOLUTION IN ACTION

The relentless toll of Darwin Awards in these chapters makes one wonder if the process of evolution has been suspended for the past twenty thousand years. If the ranks of the risk takers are continually being diminished by their own actions, why do we still have people who think it's fun to kiss a cobra? Or fly a lawnchair suspended from helium balloons into air traffic lanes?

Don't repudiate evolution, scoffing at its inefficiency! There is a good reason why it sometimes seems to be broken. Some genes linger in the pool because they have a beneficial side effect that offsets the harmful effect. Sickle Cell anemia, for instance, a well-studied genetic disease.

Sickle Cell anemia is caused by a mutation of the molecule that carries oxygen in our bloodstream. Sickle Cell hemoglobin has an abnormal sticky patch. If a child inherits Sickle Cell from both parents, all his hemoglobin is sticky. Whenever his blood is short of oxygen, such as during strenuous exercise, the sticky patches adhere to one another, causing long chains of hemoglobin to form. The chains can stretch across the entire blood cell, distorting it into the "sickle" shape that gives the disease its name. These damaged cells are fragile, carry less oxygen, and clog blood vessels. Sickle Cell anemia is an unwelcome and painful disease.

Reader Survey

How long will it take humans to evolve far enough that we run out of Darwin Award nominees?

- We're already that smart.   1%
- Another generation or two. 1%
- At least 1,000 years.   2%
- More than 10,000.   2%
- More than 100,000.   1%
- Millions of years, if ever.   8%
- **It will never happen!   84%**

Yet the incidence of Sickle Cell anemia remains high in countries where malaria is prevalent. How can such a profound genetic disease remain in our gene pool? The answer is that it actually protects "carriers" from the effects of malaria. People who inherit Sickle Cell from only one parent, instead of two, are carriers who show few adverse symptoms, yet are less susceptible to the malaria parasite.

These people are protected because their blood cells only "sickle" when the oxygen level is extremely low. And it just so happens that a cell infected with malaria has an exceptionally low oxygen level, because the parasites use up the oxygen as they multiply. This means that only malaria-infected cells are damaged, and when the cell dies, so does the malaria it was hosting.

This protective effect of Sickle Cell anemia shows how a harmful gene can also be a life-saving mutation, and helps explain why people with seemingly fatal flaws are still with us today. Perhaps being oblivious to danger, an apparently deadly trait, gave our ancestors the courage they needed to thrust a spear into a charging bison.

It may be that the incredibly poor judgment shown by the protagonists in the following stories is caused by a gene with an undetected beneficial side effect. One can only hope.

HONORABLE MENTION: LAWNCHAIR LARRY
Confirmed by Darwin
2 JULY 1982, CALIFORNIA

Larry Walters of Los Angeles is one of the few to contend for a Darwin Award and live to tell the tale. "I have fulfilled my twenty-year dream," said Walters, a former truck driver for a company that makes TV commercials. "I'm staying on the ground. I've proved the thing works."

Larry's boyhood dream was to fly. But fates conspired to keep him from his dream. He joined the Air Force, but his poor eyesight disqualified him from pilot status. After he was discharged from the armed services, he sat in his backyard watching jets fly overhead.

He hatched his weather-balloon scheme while sitting outdoors in his "extremely comfortable" Sears lawnchair. He purchased forty-five weather balloons from an Army-Navy surplus store, tied them to his tethered lawnchair dubbed the *Inspiration I*, and filled the four-foot-diameter balloons with helium. Then he strapped himself into his lawnchair with some sandwiches, Miller Lite beer, and a pellet gun.

Larry's plan was to sever the anchor and lazily float up to a height of about thirty feet above his backyard, where he would enjoy a few hours of flight before coming back down. He figured he would pop a few brews, then pop a few of the forty-five balloons when it was time to descend, and gradually lose altitude. But things didn't work out quite as Larry planned.

When his friends cut the cord anchoring the lawnchair to his Jeep, he did not float lazily up to thirty feet. Instead, he streaked into the LA sky as if shot from a cannon, pulled by a lift of forty-five helium balloons holding thirty-three cubic feet of helium each. He didn't level off at a hundred feet, nor did he level off at a thousand feet. After climbing and climbing, he leveled off at sixteen thousand feet.

At that height he felt he couldn't risk shooting any of the balloons, lest he unbalance the load and really find himself in trouble. So he stayed there, drifting with his beer and sandwiches for several hours while he considered his options. At one point he crossed the primary approach corridor of Los Angeles' LAX airspace, and Delta and Trans-World airline pilots radioed in incredulous reports of the strange sight.

Eventually he gathered the nerve to shoot a few balloons, and slowly descended through the night sky. The hanging tethers tangled and caught in a power line, blacking out a Long Beach neighborhood for twenty minutes. Larry climbed to safety, where he was arrested by waiting members of the Los Angeles Police Department. As he was led away in handcuffs, a reporter dispatched to cover the daring rescue asked him why he had done it. Larry replied nonchalantly, "A man can't just sit around."

The Federal Aviation Administration was not amused. Safety Inspector Neal Savoy said, "We know he broke some part of the Federal Aviation Act, and as soon as we decide which part it is, a charge will be filed."

Reference: Associated Press, *Los Angeles Times*, *New York Times*, UPI, Crest REACT (a C.B. radio club), STABBED WITH A WEDGE OF CHEESE by Charles Downey, ALL I REALLY NEED TO KNOW I LEARNED IN KINDERGARTEN by Robert Fulghum

Photographs of Lawnchair Larry
www.DarwinAwards.com/book/larry.html

Larry's efforts won him a $1,500 FAA fine, a prize from the Bonehead Club of Dallas, Texas, the altitude record for manned gas-filled clustered balloon flight, and a Darwin Awards Honorable Mention. He gave his aluminum lawnchair to admiring neighborhood children, abandoned his truck-driving job, and went on the lecture circuit, where he enjoyed intermittent demand as a motivational speaker. He never made much money from his innovative flight, never married, and had no children. Larry hiked into the forest and shot himself on October 6, 1993, at the age of forty-four.

DARWIN AWARD: THE LAST SUPPER
1993 Darwin Award Winner
Unconfirmed by Darwin
25 MARCH 1993

A terrible diet and room with no ventilation are being blamed for the death of a man killed by his own gas. There were no marks found on his body, but an autopsy revealed the presence of large amounts of methane dissolved in his blood.

His diet had consisted primarily of beans and cabbage, just the right combination of foods to produce a severe gas attack. It appeared that the man died in his sleep from breathing the poisonous cloud that was hanging over his bed.

Had his windows been open, the flatulence wouldn't have been fatal, but they were sealed shut to create a nearly airtight bedroom. He was an obese man with an unlimited capacity for creating methane gas, and a deadly disregard for proper ventilation.

RealAudio presentation of The Last Supper
www.DarwinAwards.com/book/realaudio6.html

Scientist Carl Sagan holds that aliens could deduce the presence of life on earth by spectral detection of methane in our atmosphere. It would prove there was continuous bioproduction of the molecule, since it breaks down rapidly in contact with oxygen. He further demonstrated that the largest source of methane is bovine flatulence, with particularly large concentrations above the cattle producing regions of the U.S. and Argentina. A single cow fart is significantly larger than a human's entire daily production.

A reader argues, "York's research on the volume and quantity of flatulence, and the fact that methane molecules break down rapidly in the presence of oxygen, make this story impossible. If we multiply the volume of air in an average bedroom by twenty-one percent for the proportion of oxygen in air, we arrive at a figure far greater than that needed to dissipate even six liters of pure methane, a value way too high for human flatulence which is in any case not one hundred percent methane.

More Flatulence Research:
www.DarwinAwards.com/book/farts.html

DARWIN AWARD:
ULTIMATE PRICE FOR SMELLING NICE
Unconfirmed by Darwin
29 JULY 1998, ENGLAND

Marcus was discovered dead by his sister Natalie in Manchester, surrounded by cans of aerosol deodorant in the bedroom of his home. Attempts to revive him failed. What killed this young man?

Apparently the aphorism Cleanliness is next to godliness was taken literally by Marcus. The seventeen-year-old reportedly bathed four times a day. His father said that Marcus doused his entire body in several kinds of deodorant at least twice a day, a routine begun six months before his death.

His parents often complained that they could "taste" the aerosols downstairs. But Marcus paid no attention to the noxious fumes or the choked out warnings. "When we told him he was using too much, he said he just wanted to smell good," his father recalled. "What a price to pay for smelling nice."

Propane and butane, the primary propellants in aerosol sprays, built up in Marcus's body during his months of excessive deodorant use. His blood contained 0.37mg/L of

RealAudio presentation of Ultimate Price for Smelling Nice
www.DarwinAwards.com/book/realaudio3.html

each toxin, nearly ten times the lethal dosage, when he suffered cardiac arrest. The coroner recorded a verdict of accidental death, citing no evidence of substance abuse. "He was simply overcome by excessive use of antiperspirants in a confined space."

Marcus's mother is calling for more prominent warnings on deodorant canisters. "We know he didn't go in for solvent abuse. He was just being meticulous about his grooming."

The British Aerosol Manufacturing Association said the death was tragic, but reiterated that aerosol deodorants are perfectly safe when used as directed.

DARWIN AWARD: MIDNIGHT SPECIAL
1992 Darwin Award Winner
Unconfirmed by Darwin
21 DECEMBER 1992, NORTH CAROLINA

Jacob, forty-seven, accidentally shot himself to death in December in Newton when, awakening to the sound of a ringing telephone beside his bed, he reached for the phone but grabbed instead his loaded Smith & Wesson .38 Special, which discharged when he drew it to his ear.

Reference: *Hickory Daily Record*

DARWIN AWARD: DEADLY READING HABITS
1993 Darwin Award Winner
Confirmed by Darwin
MARCH 1993, FLORIDA

A twenty-four-year-old salesman from Hialeah was killed near Lantana in March when his car smashed into a pole on the median strip of Interstate 95 in the middle of the afternoon. Police said the man was traveling at eighty miles per hour and, judging by the sales manual that was found open and clutched to his chest, he had been busy reading when the accident occurred.

Reference: *News of the Weird*, Universal Press Syndicate,
San Jose Mercury News

DARWIN AWARD: BREATHARIANISM
Confirmed by Darwin
22 SEPTEMBER 1999, SCOTLAND

Airhead extends principle to abdomen.

A Scottish follower of "breatharianism" demonstrated a comprehensive misunderstanding of biology when she died during an attempt to "Live with Light" in the Scottish Highlands. Verity, forty-eight, left behind a diary with references to a self-styled guru. Jasmuheen, an Australian formerly known as Ellen Greve, boasts five thousand followers worldwide, though she does not disclose whether they are always the same followers.

Verity's diary reveals that she was attempting to adhere to the twenty-one-day spiritual cleansing course wherein followers eschew all food and drink for seven days and then take only sips of water for a further fourteen days. They endeavor to master "pranic feeding" on the carbon, nitrogen, and oxygen found in the air. After that, Jasmuheen says that adherents to "breatharianism" need never eat or drink again, which she notes is the perfect cure for anorexia and world hunger.

Sufferers from anorexia and world hunger have already attempted this course of action with known results. Nutritionists say the human body can survive without fluid for no more than six days. But such research did not deter this woman, who took to the wilds with only a tent and her determination. A police source revealed that she had died from hypothermia and dehydration, aggravated by lack of food.

Jasmuheen, whose dress size was not disclosed, claims to have survived on liquid air since 1993, although she also allows herself cups of herbal tea and chocolate biscuits. In response to questioning, the founder of the cult stated that the woman's death was not due to any physical need for food; rather, it was a failure to satisfy spiritual needs brought about by a battle with her own ego.

Reference: *The Scotsman*, UK *Independent*, London *Guardian*, London *Times*, Australia *Sun Herald*

DARWIN AWARD: ROLLER COASTER
Confirmed by Darwin
1998, CALIFORNIA

An unfortunate middle-aged gentleman ignored the warning signs and suffered a particularly unfortunate fate after disembarking from the Top Gun roller-coaster at an amusement park in Santa Clara. The recipient of this Darwin Award lost his red baseball cap on the ride, and it landed under the roller-coaster. As if the danger of ground zero directly beneath a speeding roller-coaster were not self-evident, prominent RESTRICTED AREA signs dotted the perimeter fence. But after exiting the platform, the man ignored both common sense and warning signs. He climbed two fences to retrieve his hat, only to lose his head when a passenger's foot kicked his neck and derailed his plans. The woman broke her leg and lost her shoe, and is suing for damages.

References: *New York Times, Washington Post, San Jose Mercury News*

DARWIN AWARD: MENTAL ECLIPSE
Confirmed by Darwin
11 AUGUST 1999, GERMANY

A forty-two-year-old man killed himself watching the total solar eclipse while driving near Kaiserslautern. A witness saw the man weaving back and forth as he concentrated on the partially occluded sun. Suddenly he accelerated and hit the bridge pier. He had apparently just donned his solar viewers, which are dark enough to totally obscure everything except the sun.

Reference: SWR-Online.de, CNN, *Der Spiegel*, WorldOnline.nl

DARWIN AWARD:
NO BIKE LANE AT THE AIRPORT
Unconfirmed by Darwin
DECEMBER 1997, BRAZIL

A bicyclist crossing an airport runway in Sorocaba, São Paulo, was killed when he was hit by a landing airplane. Gabriel, twenty-five, could not hear the approaching plane because he was listening to his Walkman on headphones, investigators said.

Reference: Reuters

DARWIN AWARD: THE BUMBERSHOOT
Unconfirmed by Darwin
18 APRIL 1999, GERMANY

A sword-swallower died in Bonn after he put an umbrella down his throat—and accidentally pushed the button that opened it.

Reference: UK *News of the World*, KCBS News Radio

DARWIN AWARD: LEMMINGS IN A WELL
Confirmed by Darwin
20 MAY 1999, INDIA

Five people suffocated, one after another, in a particularly absurd sequence of accidents at a village well in Talaskar. A diesel pump had been employed to drain the well, and carbon monoxide and other gases from the pump filled the well. Eventually the pump ran out of oxygen and stopped working.

A youth climbed into the well to investigate the machinery, and succumbed to the choking gases and lack of oxygen. When the youth failed to reappear, another man climbed into the well and suffocated. When the second man failed to reappear, a third person climbed into the well. And so on.

Diesel fumes are not known for their beneficial effects on the human constitution. Even poorly educated people can recognize that noxious air is harmful to life and lung. But nobody considered holding his breath while investigating the faulty pump or the mounting toll of missing men.

After five people entered the well, suffocated, and died, two last would-be rescuers managed to recognize the choking feeling and rush out of the well before they, too, succumbed. They alerted police, and firemen were dispatched to the village to recover the bodies of the Darwin Award nominees.

Reference: *Indian Express*

DARWIN AWARD:
HAIR TODAY, GONE TOMORROW
Confirmed by Darwin
JANUARY 1999, ENGLAND

Some people with nervous habits have good reason to be anxious. In January a British teenager was rushed to hospital complaining of severe stomach pains. Surgeons who operated in a desperate—but ultimately unsuccessful—attempt to save her life were amazed to find a tangled mass of human hair the size of a football lodged in her abdomen.

Rachel, a seventeen-year-old hairdresser trainee, had been in the habit of chewing the ends of her tresses since early childhood. Dr. Andrew Stearman, of Poole General Hospital, Dorset, said, "The biochemical composition of hair makes it impossible for digestive juices in the stomach to break it down. It therefore accumulates, much like it builds up in the plughole of a bath or shower, attracting more hair and other food."

Recording a verdict of accidental death, coroner Alan Craze said, "This was something Rachel did from time to time by habit. She would have had the impression, if she thought about it at all, that it was passing through her system. Unfortunately, it was not, and it built to a massive size."

Pathologist Nera Patel measured the hairball—known as a "trichobezoar"—at one foot long, ten inches wide, and four inches thick. She said, "It was closely compacted and intertwined in the shape of a football. No one in our medical team had seen anything like it."

Rachel's mother, when shown a picture of the fatal obstruction, simply said, "It looks like a dead rat."

Reference: Globeandmail.com, Reuters, Wired News, the *Mirror* (UK), the *Sun* (UK), *The Mammoth Book of Tasteless Lists*

Bezoars can be made from almost anything, but are usually composed of hair. They are highly prized by shamans as protection against poison. The belief that bezoars held magical medical properties was tested in the 1500s by the barber-surgeon Ambroise Paré. He offered a convicted thief a choice between public strangulation and swallowing lethal poison along with a bezoar stone. The man chose the latter punishment, and died in agony. An affronted King Charles IX refused to relinquish his belief, and concluded that the bezoar stone was a fake.

DARWIN AWARD: DEATH OF DRACULA
Unconfirmed by Darwin

A college student costumed himself as Dracula for Halloween. As a finishing touch he put a pine board down the front of his shirt so he could "realistically" sink a knife into the board and pretend he was transfixed by a vampire-killing stake. He didn't consider the strength of the thin pine board when he tapped the knife in with a hammer. Propelled by the force of the hammer, the sharp blade split the soft pine and buried itself in his heart. He staggered from his dormitory room into the Halloween party, gasping, "I really did it!" before succumbing before horrified friends.

Reference: *Dead Men Do Tell Tales* by William R. Maples, Ph.D., 1994

DARWIN AWARD: THE DAILY GRIND
Confirmed by Darwin
1 MARCH 2000, MAINE

He really got caught up in his work.
The owner of Carrier Chipping, Inc., inadvertently reproduced the chilling climactic scene in the movie *Fargo*, and was rent asunder by his own wood chipper.

The chipper that did him in is affectionately known as the Hog. It will take birch or maple logs up to twenty-four inches in diameter and reduce them to three-quarter-inch chips of wood.

Employees were working late to make up for time spent repairing equipment malfunctions earlier in the day. When the Hog jammed, Michael climbed the conveyor belt feeding the chipper and used a rake to break up the bark jam in the chute.

Director C. William Freeman of the Bangor Occupational Safety and Health Administration said, "Generally, our experience (of fatal accidents involving chippers) has found two causes: inadequate machine guarding, or a failure to institute an effective lockout-tagout program when someone is unjamming pieces of equipment." Apparently Michael was not a proponent of lockout-tagout procedures. His unjamming efforts were directed against a machine that was still in operation.

The Skowhegan resident was somewhat the worse for wear after his passage through the Hog. Police Chief Butch Asselin said that the remains would be subjected to DNA analysis for a positive ID, and added "I hope I never, ever see anything like this again."

Reference: Blethen Maine Newspapers, *Kennebec Journal*

URBAN LEGEND: CHRISTMAS ROAST
25 DECEMBER 1998, CANADA

Telephone relay company night watchman Henry Baker, thirty-one, was killed early Christmas morning by microwave radiation exposure. He was apparently attempting to keep warm next to a telecommunications feedhorn.

Henry had been suspended on a safety violation once before, according to Northern Manitoba Signal Relay spokesperson Tanya Cooke. She reported that Henry's earlier infraction was for defeating a safety shutoff switch and entering a restricted maintenance catwalk in order to warm himself in front of the microwave dish. He told coworkers that it was the only way he could stay warm during his twelve-hour shift at the station, where winter temperatures often dip to forty below zero.

Microwaves can heat water molecules in human tissue the same way they heat food in a microwave oven. For his Christmas shift Henry reportedly brought a twelve-pack of beer and a plastic lawnchair, which he positioned directly in line with the strongest microwave beam. Henry was unaware of the tenfold boost in microwave power planned that night to handle the anticipated increase in holiday long-distance calling traffic.

Henry's body was discovered by the daytime watchman, John Burns, who was greeted by an odor he mistook for a Christmas roast he thought Henry must have prepared as a surprise. Burns reported to NMSR Company officials that Henry's unfinished beers had also exploded.

Reference: Hoax perpetrated by NMSR
(New Mexicans for Science and Reason)

URBAN LEGEND:
THE LAUNDRY WAS CLEAN. . . .
1998

A thirty-nine-year-old Charlottesville man died in a freak accident involving his washing machine. According to police reports Ned Hurt was doing laundry when he took extraordinary measures to speed up the process.

Ned apparently tried to stuff fifty pounds of laundry into his washing machine by climbing on top of the washer and forcing the clothing into the basin. Ned then accidentally kicked the washing machine's on button. When the machine began to fill with water, Ned lost his balance and both feet slipped down into the basket, where they stuck fast.

The machine started its cycle, and Ned, unable to free himself, began thrashing around as the agitator went into gear. Ned's head banged against a nearby shelf in the laundry room, knocking over a bottle of bleach that poured over his face, blinding him. Forensic reports say Ned also swallowed some of the bleach and vomited. He was still unable to free himself from the bowels of the washer.

The smell of the vomit attracted Ned's dog into the laundry room. At the same time, according to police, a large box of baking soda fell from the shelf, startling the dog, who urinated. Urine, like vinegar, is acidic, and the chemical reaction between the urine and the baking soda resulted in "a small explosion," according to police reports.

Ned remained stuck in the washing machine, which

eventually went into its high-speed spin cycle, spinning Ned around at seventy miles per hour, according to forensic experts. Ned's head smashed against a steel beam behind the washing machine, immediately killing him. A neighbor heard the commotion and called 911, but by then it was too late. Ned was pronounced dead at the scene.

The dog escaped unharmed.

URBAN LEGEND: UNFORTUNATE HUSBAND

This legend comes in two versions, set a year apart.

1996, CALIFORNIA

A Los Angeles husband was deathly afraid of heights. Nevertheless, one day he found it necessary to climb onto his roof to adjust the TV antenna. His fear impelled him to take precautions against falling from the roof. He tied a sturdy rope around himself and affixed the other end to the bumper of his car. Unfortunately, he neglected to inform his wife of his activities.

She had just finished making a shopping list. She climbed into the car and set off for the store, pulling her husband from the roof and dragging him a quarter mile down the street before a startled neighbor alerted her to her extra cargo. The man was rushed to the hospital, where he spent many days recovering from broken ribs and severe lacerations.

The story does not end there. To make amends the contrite wife planned a surprise party for her husband on the day of his homecoming. She invited several mutual friends over, most of them smokers. Since the wife and husband also smoked, they had several lighters around the house. The wife was a good hostess, and she decided to fill them over the toilet before the guests arrived.

Can you guess what happened?

The husband needed to use the bathroom immediately

after he got home. He sat on the toilet, picked up a magazine, and threw his cigarette into the toilet. . . .

Kaboom!

1997, CALIFORNIA

A San Francisco resident pushed his motorcycle from the patio into his living room, where he began to clean the engine with some rags and a bowl of gasoline. Much better to clean the bike in the comfort of his own home, he thought, than out in the chilly San Francisco fog. When he finished, he sat on the motorcycle and decided to give his bike a quick start and make sure everything was still okay. Unfortunately, he started up the bike in gear, and it crashed through the glass patio door with its rider still clinging to the handlebars.

His wife had been working in the kitchen. She came running at the fearful sound, and found him crumpled on the patio, badly cut from the shards of broken glass. She called 911, and the paramedics carried the unfortunate man to the emergency room.

Later that afternoon, after many stitches had pulled her husband back together, the wife brought him home and put him to bed. She cleaned up the mess in the living room and dumped the bowl of gasoline in the toilet.

Can you guess what happened next?

Shortly thereafter her husband woke up, lit a cigarette, and went into the bathroom for a much-needed relief break. He sat down and tossed the cigarette into the toilet, which promptly exploded because the wife had not flushed the gasoline away. The explosion blew the man through the bathroom door.

The story does not end there. The wife heard a loud explosion and the hideous sound of her husband's screams. She ran into the hall and found her husband lying on the floor with his trousers blown away and burns on his buttocks. The wife again ran to the phone and called for an ambulance.

The same two paramedics were dispatched to the scene. They loaded the husband on the stretcher and began carrying him to the street. One of them asked the wife how the injury had occurred. When she told them, the two paramedics began laughing so hard that they dropped the stretcher and broke the man's collarbone.

URBAN LEGEND: OVERKILL
This story has two parts, one apocryphal and one true.
1999, FRANCE

Pierre Murard left nothing to chance when he decided to commit suicide. He stood atop a sheer cliff and tied a noose around his neck. He tied the other end of the rope to a large rock. He drank some poison and set fire to his clothes. He even tried to shoot himself at the last moment, firing the pistol as he jumped from the cliff.

The bullet missed him completely and cut through the rope above him. Now freed from the threat of hanging, he plunged into the sea. The dunking extinguished the flames and made him vomit the poison. He was dragged out of the water by a kind fisherman and taken to a hospital, where he died of hypothermia.

Confirmed by Darwin
28 JULY 1999, MADRID

A Spanish man who made an unsuccessful bid for a Darwin Award was recovering in the hospital after two failed suicide attempts. The despondent thirty-eight-year-old man had leapt from his third-floor apartment, but failed to even knock himself out when he hit the ground. The unhappy man staggered back into his apartment and cut his throat, but police broke down the door and took him to a hospital in the town of Mérida. His wound was stitched up and he was released after counseling for depression.

Reference: Reuters, Europa Press, *Houston Chronicle*

PERSONAL ACCOUNT: TEAM SPIRIT
1998, FLORIDA

Florida is stuffed with staunch baseball fanatics. A few days before the big Marlins game, two supporters decided to show their loyalty by constructing an enormous paper Marlins banner. Naturally they wanted to display their efforts prominently. They chose to drape it across the Metro Rail overpass where it crossed a major thoroughfare. The banner would proclaim the superiority! not to mention the high testosterone content! of the city!

Unfortunately, their plans were not altogether complete. Neither had the foresight to procure a Metro Rail schedule. Lacking essential knowledge of the train timetable, they hung the banner just as the automated Metro thundered implacably toward them and struck them both. One was killed, the other left wounded to tell the tale.

The banner was reportedly unharmed.

Reference: Anonymous personal account.

PERSONAL ACCOUNT: CLEANING THE HEAD
1999, WASHINGTON

There comes a time in a man's life when he is beset by the will to clean the chemical toilet in his RV. Unfortunately, in this man's life the old OpenRoad RV had not been used in many years, and the toilet was nearly solid with waste.

He pondered the problem, and realized that he needed a suitable liquid to pour into the toilet to unplug the muck and get it flushing again. In a stroke of brilliance he promptly poured in a bottle of that little-known drain opener, bleach. It immediately became apparent that this was not the right move, as an expanding cloud of chlorine gas began to fill the RV.

Choking sounds filled the air as the poor sap struggled with the antiquated back door latch and finally managed, just before lapsing into unconsciousness, to open it. He fell gasping headlong into the outside air.

He survived to learn a valuable lesson about how body wastes, especially urine, decompose into ammonia, thus supplying a reagent perfectly suited to react with the bleach and emit a deadly cloud of gas.

Reference: David "Who, Me?" Brager, personal account.

Appendices

I. Website Biography

The Darwin Awards archive was born on a Stanford University webserver in 1994. It spun off a regular email newsletter that is still available today. News of the website spread, and nominations culled by avid followers flew in from far and wide. As the number of stories in the archive grew, so did its acclaim. It became the locus for new Darwin Awards.

Eventually the website became more popular than Stanford's research-oriented webpages. The sysadmin suggested that it might be happier on someone else's server, and www.DarwinAwards.com was born.

www.DarwinAwards.com is the principal repository for official Darwin Awards and associated tales of misadventure. New nominees are submitted daily, and made available for public comment in the Slush Pile. The website nurtures a community of free thinkers who debate the merits of controversial nominees in the Philosophy Forum. The Forum is also home to a wide array of philosophical, political, and artistic conversations. Visitors to the website can sign up for a free newsletter, and register to win Darwin Awards T-shirts and other paraphernalia.

Stories in this book may include a URL directing you to a webpage with more information. All of the hyperlinks in this book can be explored if you start at this portal:

Darwin Awards: Evolution in Action
www.DarwinAwards.com/book/

The Darwin website has attracted compliments from a diverse crowd of celebrities, from sexy Miss Shea Marks to wry David Boze and all points in between.

- "Thank goodness for the authentic manual for things *not* to do if you want to grow old."
  —Chuck Shepherd, *News of the Weird*
- "Good, harmless fun—harmless for the reader, anyway! Darwin Awards break up the monotony of the day."
  —Shae Marks, Miss May 1994
- "Only a monkey wouldn't love the Darwin Awards. Tickles the funnybone even as it shakes your faith in mankind."
  —David Boze, Producer, *The Weissbach Show*
- "*USA Today* Hot Site: The latest, ahem, winners ..."
  —Sam Meddis, Technology Editor, *USA Today*
- "One of the Internet's most bizarre cult followings—the Darwin Awards is a great example of the community and dynamism of the Internet."
  —Libby Jeffery, *Web Wonders*
- "One of those guilty little pleasures, like sitting down with a pint of ice cream and watching trashy TV."
  —Jules Allen, *St. Petersburg Times* Site Seeing Pick
- "With the aid of Darwin Award winners, the human race can hope to move boldly forward along the evolutionary path. Dead brilliant!"
  —Danielle Buhagiar and Meagan Loader, Kgrind Live Online Radio
- "One of the funniest, most inventive, and downright interesting websites I have come across."
  —Greg Smith, Cool Site of the Year

II. Darwin Haiku

You've heard of "spam haiku." Now meet "darwin haiku" from the Philosophy Forum.

Hat on railroad track
Leave it unless you hate your
own appendages.
Dev

I think for myself.
The warning signs do not apply
to an immortal.
haiku

Homemade parachute,
not latest technology,
opens on impact
Jericho

Death happens to all.
Some have quite a while to
wait.
Some knock on Death's door.
Tardis

Stupidity dies.
The end of future offspring.
Evolution wins.
haiku

The dawn of Mankind.
When will our journey be
through?
Dusk could be here soon.
a3dsail

Stupidity kills.
Absolute stupidity.
Kills absolutely.
Silverhill

Dodo is extinct.
Rhino's sitting on the brink.
You're next, save you think.
ddiggler

Canada in white.
Snowmobiler's paradise.
Darwin's gonna strike!
ddiggler

Dude had a screw loose.
Glad he didn't reproduce.
Darwin saved the day.
ddiggler

III. Author Biography

Wendy Northcutt studied molecular biology at UC Berkeley, then worked in a neuroscience research lab at Stanford University. She launched the Darwin Awards archive on a Stanford website, and emailed stories to a small list of friends. When academia began to pall, she joined a start-up company hoping to develop cancer and diabetes therapeutics, and continued to work on the Darwin Awards.

Eventually the lure of the Internet proved too strong to resist, and Wendy abdicated her laboratory responsibilities to pursue a dream. She now works as a freelance webmaster, and hones her skills on the Darwin Awards website. Today, nominations from a worldwide network of fans are presented for vote and debate at www.DarwinAwards.com.

Wendy first learned of the concept of the Darwin Awards from her cousin Ian, a mildly eccentric philosopher who later started his own religion in order to avoid shaving his beard while working in the pizza industry. Ian is now pursuing a degree in archeology, and his hair is no longer an issue.

Wendy devotes her free time to studying human behavior, writing Darwin Awards, reading, traveling, and gardening.

Story Index

WELCOME TO THE NEXT EVOLUTION IN HUMOR.

BACK BY POPULAR DEMAND!

In the *New York Times* bestseller, *THE DARWIN AWARDS II: Unnatural Selection*, author Wendy Northcutt returns with a fresh collection of cautionary tales and serious humor.

NOW AVAILABLE IN DUTTON HARDCOVER